INDUSTRIAL HYGIENE MADE EASY

A Checklist Approach to Recognizing, Evaluating, and Controlling Workplace Hazards

Darolyn K. Wall, CIH, CSP

Foreword by
S. Zack Mansdorf, PhD, CIH, CSP, QEP

MORAN ASSOCIATES
Orange Park, FL • Washington, DC

Acquisitions Editor: Alexander M. Padro
Production Editor: Terrence E. Fischer
Designer: Pleasant P. Mann
Cover Designer: Alexander M. Padro

Moran Associates
1600 Brighton Bluff Court
Orange Park, FL 32073-7409

Phone: 904-278-5155
Toll-Free: 1-800-597-2040
Fax: 904-278-5494
E-mail: sales@moranbooks.com
Web Site: http://www.moranbooks.com

First Edition ISBN 1-890966-51-7

Printed in the United States of America

03 02 01 00 99 5 4 3 2 1

Library of Congress Catalog Card Number: 99-066701

Table of Contents

Foreword

The art and science of industrial hygiene progressed significantly in the 20th century. This development was driven by a wide variety of factors. These include economic, social, and moral imperatives coupled with the advent of OSHA and a vastly improved understanding of occupationally related disease. This increase in knowledge and the enormous breadth of the field can present a formable challenge for someone wanting to master its many diverse aspects and broad applications.

This challenge has been made even more difficult by the changes in company and corporate structures during the past decade. Responsibility for the identification and resolution of safety, health and environmental issues has been increasingly placed on operating units and work teams, many times without the aid of technical support from a large corporate headquarters.

Industrial Hygiene Made Easy will help meet the challenge. It is a very well written, comprehensive and easy to follow reference for both the aspiring and practicing industrial hygienist or someone who simply wants to learn about a particular subject. Darolyn Wall has taken a unique approach to our thirst for concise information presented in easy to read snippets and "factoids." While the arrangement of the text is in short, concise paragraphs, the information presented is as broad and valuable as that in larger texts on the subject. As a measure of the comprehensive nature of the book, it has 20 chapters covering everything from recognition of potential health hazards to their control.

It also goes beyond the usual topical coverage for a fundamental text in industrial hygiene. This includes valuable chapters on topics such as the historical development of the profession; biological hazards, ergonomics, industrial processes and their associated health hazards, environmental issues, and ethics and management. Those new to the field will especially appreciate the chapter on industrial processes and associated health hazards and the chapter on chemical hazards.

Both the aspiring industrial hygienist and the seasoned practitioner will find this a very useful text for their bookshelf. I join the author in the hope that those of you reading and using this book will find industrial hygiene a very rewarding and desirable career. I am confident that your employers will benefit from your knowledge in protecting our most valuable resource: workers.

S. Zack Mansdorf, PhD, CIH, CSP, QEP
Beverly, MA

Preface

Industrial Hygiene Made Easy is intended for use by students, physicians, nurses, consultants, safety professionals, and environmental professionals in industry, government agencies, labor organizations, and consulting practices who must gain additional skills in recognizing, evaluating, and controlling occupational health hazards. It is ideally suited for use by busy professionals who need a refresher or must quickly learn and apply industrial hygiene principles in the workplace.

In this era of downsizing and restructuring, health, safety, and environmental professionals are finding themselves in less functional or specialist positions and cast into generalist roles. Many large and small corporations have successfully downsized and restructured their organizations to operate leaner and more competitively. Some downsized organizations have outsourced health, safety, and environmental services. (Outsourcing is a commonly-used term for using contract employees or consultants to perform jobs that are not consistent with a company's core product or service.) Other downsized organizations expect the remaining employees to learn the duties of the departed employees. These survivors must expand from the specialist role to that of the generalist.

With adequate preparation, becoming generalists can be an excellent method for health, safety, and environmental professionals to add more value to their organizations. However, a study by Right Associates, an international career management and human resources consulting firm, found that 88 percent of the survivors of downsizing were given no training or assistance in how to function more effectively in the new environment. If true, physicians, nurses, safety, and environmental professionals forced to expand into the industrial hygiene arena may feel intimidated, even frightened, by the prospect that they must perform complicated tasks without adequate preparation. Even students focusing on safety degrees and environmental degrees are finding that companies want employees with broader skill bases.

Industrial hygiene traditionally has been viewed as a highly scientific and engineering based profession. Without preparation or training, a transition into industrial hygiene can be difficult. If you must expand your skills into the industrial hygiene field or just want to add more value to your company, *Industrial Hygiene Made Easy* covers the essential areas of industrial hygiene in a fundamental manner.

Industrial Hygiene Made Easy is intended for use as a self-study text or as an instructional text for a fundamental industrial hygiene course. The book is both an easy-access aid to the implementation of an industrial hygiene program by safety and environmental professionals, and a ready reference for improving the health and safety of the workplace. *Industrial Hygiene Made Easy* promises to make learning easy.

This book was written in a checklist style that features headings, block labels, and block text. This format permits readers to quickly scan the block labels on the left-hand side of the page and refer to the points of interest. The block format also eliminates transitional sentences, which allows readers to return easily to find answers without having to cull through lengthy, unbroken text.

Industrial Hygiene Made Easy introduces readers to the history of industrial hygiene and its evolution into a profession. The principles of recognition, evaluation, and control of common workplace hazards are presented. Chemical and biological hazards are examined. Physical hazards are covered in chapters on ionizing radiation, non-ionizing radiation, noise and noise control, and ergonomics. The toxicology chapter highlights specific chemicals and metals and describes their unique hazards. The gas laws and their significance in calculating air sample results are emphasized. Air sampling is covered in chapters on gases and vapors and direct-reading instruments. There is also a chapter describing the significance of particulate sampling and analysis. The chapter on industrial hygiene statistics explains sampling and statistical errors.

In today's competitive business climate, we must earn our positions every day. We must enable our organizations to get more out of us than in the past. In other words, we must make sure we add more value than we cost. What counts is our contribution, not how busy we are, how many hours we clock in, how much face-time we put in—not even how many years we put in. Long gone are the days when organizations promoted someone based on longevity alone. It is my intention that readers not only learn new skills from this text, but that they become energized to apply these new skills and stand out among their peers.

Darolyn K. Wall, CIH, CSP
Chandler, AZ

About the Author

Darolyn K. Wall, director of Scarlett Professional Development, offers health and safety consulting to a wide range of industries and corporations. Her specialty is organizational development, helping health, safety and environmental professionals to better integrate their responsibilities and functions into their overall businesses.

As a corporate insider, Wall has held various positions, including 10 years with a Fortune 500 company. She began her career as an industrial hygienist. Her strong technical background and leadership abilities earned her several management positions in the aerospace industry.

Wall earned a Master of Science degree in Preventive Medicine and Environmental Health from the University of Iowa. She is as a Certified Industrial Hygienist (American Board of Industrial Hygiene) and a Certified Safety Professional (Board of Certified Safety Professionals).

Wall is a member of the American Industrial Hygiene Association and the American Society of Safety Engineers. She is an advisor to the Board of Directors of the Arizona Occupational Health Nurses Association, and serves her community as a public education volunteer for the Chandler, AZ Fire Department. Wall is listed in the *International Who's Who of Professionals*.

1 Introduction to Industrial Hygiene

Industrial hygiene is the science and art dedicated to the anticipation, recognition, evaluation, and control of environmental factors or stresses arising in or from the workplace. Such workplace environmental stresses may cause illness, impaired health, or significant discomfort among workers or the citizens of a community.

History of Industrial Hygiene

INDUSTRIAL HYGIENE'S ORIGINS

- ❑ Industrial hygiene originated with physicians identifying adverse health effects in their patients and subsequently associating these illnesses with specific occupations.
- ❑ Industrial hygiene has progressed from simply recognizing occupational illnesses with very little done to protect workers, to basic hygiene, to worker protection, and finally to legislation.

4th Century BC to 1st Century AD

- ❑ Lead toxicity in the mining industry was recognized by Hippocrates, although protection of the worker was not a concern at the time.
- ❑ Pliny the Elder, a Roman scholar, identified the dangers inherent in dealing with zinc and sulfur. He described a bladder-derived mask for protecting laborers from dust or lead fumes.

2nd Century to 15th Century

- ❑ In the second century, Galen, a Greek physician, identified the dangers to copper miners posed by acid mists.
- ❑ In 1473, Ulrich Ellenbog wrote a notable pamphlet on occupational diseases and highlighted hygiene instructions.

16th Century

- ❑ In his book *De Re Metallica*, Georgius Agricola suggested mine ventilation and protective masks for miners.
- ❑ Effects on the extremities due to lengthy exposure to the cold water of damp mines (trench foot) and a disease of the lungs caused by inhalation of silica or quartz dust (silicosis) were first recognized around 1556 by Agricola.
- ❑ In 1580, Paracelsus, a Swiss doctor, introduced remedies and medicines for diseases associated with miners. Although quite advanced, his conclusions related to mercury poisoning and toxicity, sulfur and salt vapors were erroneous.

18th Century

- *De Morbis Artificum*, written in 1700 by Bernardo Ramazzini, an Italian physician, had a tremendous effect on the future of public hygiene. Ramazzini was the first to ask each patient the question that was later to be included in almost every physician's case history: "Of what trade are you?"
- Physician Sir George Baker accurately attributed Devonshire Colic to lead in the cider industry and contributed to its removal.
- Percival Pott, a physician renowned for recognizing soot as one of the causes of scrotal cancer, was instrumental in the passage of the Chimney Sweeper's Act of 1788.

19th Century

- The English Factory Act of 1833, the first effective legislation to require attention be given to the working population, was in practice more directed toward providing compensation for accidents than controlling the causes of these accidents.
- Other European countries soon developed Worker's Compensation Acts.
- A sense of community responsibility evolved, heightened by the interest of newspapers in the control of the environment.
- New laws stimulated the adoption of increased factory safety precautions and introduced medical service in industrial plants.
- The last of the English factory acts established a centralized inspection office in London charged with conducting factory inspections.

20th Century

- The United States' early 20th century workplace health champion was physician Alice Hamilton. Dr. Hamilton presented substantiated evidence of a relationship between illness and exposure to toxins, and went further by suggesting solid solutions to these problems. Subsequently, US public awareness increased and legislation began in this country.

20TH CENTURY LEGISLATION IN THE UNITED STATES

- In 1908, the federal government passed a compensation act for civil employees.
- The first state compensation laws were passed in 1911, and by 1948 all states had passed compensation laws. State compensation laws were a fundamental factor in the development of industrial hygiene in the US, as it became more economical to control hazards than pay compensation.

- In 1910, the US Bureau of Mines was the first federal agency to conduct studies in the mining and steel industries.
- The US Department of Labor, and two states, New York, and Ohio, initiated industrial hygiene programs in 1913.
- Between 1912 and 1915, the US Public Health Service embarked on investigations in many industries and developed workable solutions.
- The Metal and Nonmetallic Mine Safety Act was passed in 1966 detailing health and safety standards for mines.
- The Federal Coal-Mine Health and Safety Act of 1969 attempted to attain the highest degree of health protection for miners.
- The Occupational Safety and Health Act of 1970 created the Occupational Safety and Health Administration (OSHA), assuring so far as possible every working man and woman in the nation safe and healthful working conditions and preserving human resources.

INDUSTRIAL HYGIENE AS A PROFESSION

- Industrial hygiene has evolved from an abstract and undefined science to a recognized and respected profession. Industrial hygienists are professionals with the training, education, and experience in anticipating, and recognizing health hazards arising from work processes and operations.
- Industrial hygienists evaluate work-related environmental hazards and stresses through measurement of chemical, physical, ergonomic, and biological stresses. Skilled at evaluating and measuring the magnitude of hazards, they are equally equipped to establish and implement effective hazard control methods for reducing or eliminating exposure.

Recognition of Potential Health Hazards

RECOGNITION

- Industrial hygienists follow general principles to assist with their recognition of potential threats to workers.
- Industrial hygienists must:
  - Conduct research and extensive field study to become familiar with the processes, job activities performed, and effective control measures in place, to determine hazard potential.
  - Become familiar with all work processes used in the particular workplace under scrutiny.

- Learn what chemicals or materials are used, and what intermediate products, or byproducts are produced.
- Physically survey the workplace under evaluation, ask questions during the survey, and collect process flow diagrams and other information on the layout of the facilities.
- Prepare and maintain an inventory of the chemical and physical agents used in the workplace.

DETERMINING IF A CHEMICAL IS TOXIC

- Industrial hygienists research and review reference books and material safety data sheets for specific information on chemicals or products used in the workplace to determine which of these are toxic and to what degree.

- Information about the toxicity of chemicals can be found in the latest texts and scientific journals. Chemical dictionaries provide the concentration at which each substance is known to be toxic, and how the chemical is formed and used. They may identify the threshold limit value (TLV) outlined by the American Conference of Governmental Industrial Hygienists (ACGIH) or the permissible exposure limit (PEL) mandated by the Occupational Safety and Health Administration (OSHA).

Toxicity Versus Hazard

- Industrial hygienists recognize that toxicity and hazard are not synonymous.

- The injury produced when a chemical has reached a sufficient concentration in the body is referred to as toxicity.

- The hazard of a chemical is the probability that this concentration will be reached within the body.

- The nature of the work process in which a potentially hazardous substance is used, its possible reaction with other materials, and the degree of effective control measures all affect the potential hazard an agent may represent.

- The severity of a hazard depends on many variables an industrial hygienist must consider, including how a chemical is to be used, the type of job performed, the duration of exposure, and work patterns.

- An industrial hygienist must review several factors.
  - By what route of entry will the chemical enter the body?
  - How much of a chemical and for how long must a chemical be in contact with cellular components of the body?
  - What is the likelihood that the material will be absorbed by the body and come in contact with body cells?

- What is the generation rate of the airborne contaminants?
- What effective control measures are in place or must be in place to minimize adverse health effects?

PRELIMINARY SURVEY

- Industrial hygienists conduct a preliminary survey of the work area.
- As a preliminary survey is conducted, many potentially hazardous work processes can be visually observed.
- Operations that generate fumes (such as welding) are easily recognized as a potential hazard and quickly indicate that further evaluation may be required.
- Operations where chemicals are used for cleaning or degreasing work-in-process (WIP) are also easily recognized and indicate that further evaluation is necessary.
- The presence of vapors or gases may be detected by the sense of smell.
- Review workers' routine job requirements.
- Visually observe how work is performed and whether all workers perform their jobs in the same manner, or if worker-to-worker variability may affect their exposures.
- Review the types of control measures in use, and examine their effectiveness. These include local exhaust ventilation, personal protective clothing or equipment, shielding from ultraviolet radiation, etc.
- Take notes during the preliminary survey, draw layouts of the work area, note whether housekeeping is adequate, whether workers eat or drink in the work area, if control measures are adequate but not used, etc.

Typical Investigation Questions

- Typical questions that should be asked during a preliminary survey include the following:
  - What is produced? What raw materials are used?
  - Are any of these raw materials regulated by OSHA or the Environmental Protection Agency (EPA) and require employee exposures to remain within permissible exposure limits (PELs)?
  - What materials are added? Are materials generated?
  - What equipment is involved?
  - What operational procedures are involved? Do these procedures minimize exposures?
  - Is there a written procedure for the safe handling and storage of materials?

- What about dust control, cleanup after spills, and waste disposal?
- Is there a ventilation or exhaust system in place? If so, is it adequate?
- Does the facility layout minimize exposure?
- Is the facility well equipped with safety appliances such as showers, masks, respirators and emergency equipment?
- Are safe operating procedures outlined and enforced?
- Do these include procedures for maintenance personnel and non-routine operations?
- How many work-shifts are there? How many workers perform these job functions?
- Do workers leave the area during their lunch break?
- Do workers exceed 40 hours a week with extensive overtime?
- Are all workers skilled and adequately trained?

SOURCES OF AIRBORNE CONTAMINANTS

- The list of raw materials, products and byproducts used and generated, and the results of the preliminary survey, will indicate the source and type of potential airborne contaminants.

- The various forms of airborne contaminants and their sources are outlined below.
  - Aerosol: Suspension of liquid or solid particles in air.
  - Dust: Particulate material generated by a mechanical process (0.5 to 50µm).
  - Mist: Suspension of liquid particles in air formed by condensation from vapor or by some mechanical processes (40 to 400 µm). For comparison, rain = 100µm.
  - Fume: Solid particle aerosol formed by condensation from the vapor state (0.001 to 0.2 µm).
  - Fiber: Particle with aspect ratio (length/width) greater than 3:1 (NIOSH 5:1).
  - Smoke: Aerosol formed from combustion of organic material (0.01 to 0.5 µm).

ROUTES OF ENTRY

- In order for a toxic agent to exert its effect, the agent must first come in contact with a worker's body. This can occur through inhalation (breathing), absorption (through direct contact with skin), or ingestion (eating or drinking).

Inhalation

- Inhalation involves airborne contaminants that are breathed into the lungs. Airborne contaminants include gases, vapors, and particulate matter (including dusts, fumes, smokes, aerosols, and mists).

- Inhalation is the most rapid form of entry for a contaminant to enter the body. The airborne contaminant is inhaled, absorbed in the lungs, picked up by the bloodstream and distributed throughout the body.

Absorption

- Absorption of a chemical through the skin is a much slower method of entry, but occurs very quickly if the skin is cut or damaged.
- Many compounds can be absorbed directly through the skin. Phenols, benzene and organic lead compounds are examples of materials that can be absorbed through the skin.

Ingestion

- Failure to follow proper hygiene practices may allow chemicals to enter the body through ingestion.
- Toxic compounds can then be absorbed into the blood from the gastrointestinal tract. Lead oxide is an example of a material that can cause serious physical harm if ingested.

Evaluation of Potential Health Hazards

EVALUATION CRITERIA

- The work process must be understood in order to identify where contaminants are released.
- Extensive information must be obtained regarding:
  - Types of hazardous materials used in a facility.
  - The kind of jobs performed.
  - How the workers are exposed.
  - Work patterns and how they vary between employees.
  - Levels of air contamination.
  - Duration of exposure
  - Control measures used.
  - Other necessary information.

Information Required

- For each process, the following must be performed:
  - Observe who performs work, how, where, how long, and by how many people it is done.
  - Observe how health hazards are generated, and the control measures in use.
  - For each contaminant, find the OSHA PEL or other recommended exposure guideline based on the toxicological effects of the material.
  - Determine the actual level of exposure to harmful agents.
  - Determine the number of employees exposed and length of exposure.
  - Identify the chemicals and contaminants in the process.

- ❑ Determine the level of airborne contaminants using air sampling or other analytical techniques.
- ❑ Calculate the resulting daily average and peak exposures from the air sampling results and employee exposure times.
- ❑ Compare the calculated exposure with OSHA standards, the TLV listing from the ACGIH, or other toxicological recommendations.

AIR SAMPLING

❑ To effectively evaluate a potentially hazardous workplace, objective and quantitative analysis is necessary. This usually requires some form of air sampling.

❑ Air sampling instruments are categorized as follows:
- ❑ Direct reading.
- ❑ Those that remove the contaminant from a measured quantity of air.
- ❑ Those that collect a known volume of air for subsequent laboratory use.

Instrument Selection

❑ Instrument selection depends on various factors including:
- ❑ Type of contaminant under evaluation.
- ❑ Instantaneous results or whether laboratory analysis is required.
- ❑ Portability.
- ❑ Ease of use.
- ❑ Efficiency of the device.
- ❑ Reliability of the equipment.
- ❑ Type of analysis.
- ❑ Availability of the equipment.

Sensitivity, Accuracy, and Reproducibility

❑ When evaluating a worker's exposure or the environment, an instrument must be used that can provide sensitivity, accuracy, and reproducibility.

❑ Factors to consider when selecting equipment are:
- ❑ Whether or not the particular instrument is specific for the contaminant to be determined.
- ❑ What other substances interfere with the test.
- ❑ The accuracy and sensitivity of the device.

Equipment Calibration

❑ Calibration procedures must be used to insure that analysis of field samples accurately represents concentrations in the environment, and specifically, concentrations to which a worker is exposed.

❑ It is critical that air-sampling equipment operate at a known rate of airflow. The equipment must be calibrated against a standard airflow measuring device both before and after use in the field.

- The precise rate of airflow must be recorded so that when it is multiplied by the sampling time, the total volume of air sampled or collected will be known. This volume of air is used in calculating the concentration of contaminant to which the worker was exposed.

Standard Methods

- Standard methods of analysis should always be used. The National Institute of Safety and Health (NIOSH) has developed a *Manual of Analytical Methods* that contains many established and recognized methods of collection and analysis. When in doubt, many industrial hygienists work very closely with the laboratory that will be performing the tests to assure that proper sample collection, storage and handling methods are utilized.

THE FIELD SURVEY

- Every effort must be made to obtain samples that represent the worker's exposure.

- To determine what constitutes a representative sample, an industrial hygienist must decide:
  - Where to sample.
  - Whom to sample.
  - How long to sample.
  - How many samples to take.
  - When to sample.

Where to Sample

- Where the purposes of sampling are to determine a worker's exposure and daily time-weighted-average, it is necessary to collect samples at or near the breathing zone.

- Actual exposure levels can be documented by combining the proper number of personnel samples with the results from stationary sampling points.

Whom to Sample

- Monitoring different job tasks in a suspect area is best accomplished by personnel exposure monitoring.

- Properly attaching personal monitors to workers directly exposed should provide a representative sample of actual breathing zone exposure.

How Many Samples to Take

- No set rule exists to determine the number of samples that are necessary to evaluate a worker's exposure.

- It is necessary to take enough samples to truly characterize the exposure for the particular job. A single sample is never enough.

- Samples should be taken for a number of workers over several days to reflect a typical exposure.

Duration of Sampling

- When determining the volume of air to sample and the duration of sampling, it is necessary to consider:
  - Sensitivity of the analytical procedure.
  - Estimated air concentration.
  - The TLV of the particular contaminant.

- The best results can be obtained (when analytical methods will permit) by allowing the worker to work a full eight-hour shift with a personal breathing zone sampler attached for the duration of that shift.

- Full-shift integrated personal sampling is preferable to that of short term or general area sampling if the results are to be compared to standards based on time-weighted average concentrations.

When to Sample

- Sampling strategies should be repeated every few months to determine if seasonal or production schedule variations affect exposure. It has been found that individual exposures may be different for each shift.

- If multiple shifts are used, sampling employees on each shift should be considered.

- If production schedules are non-linear, sampling at the end of the month or during peak production periods may also affect personnel exposures, therefore sampling during low and high production schedules is critical to assessing accurate exposure levels.

INTERPRETATION OF RESULTS

- The final step in evaluating the workplace environment is the interpretation of analyses.

- Prior to determining that a group of workers has been exposed to an injurious hazard, the following must be established:
  - Nature of substance or physical agent involved.
  - Concentration of exposure.
  - Duration of exposure.

- Air sampling results are expressed as time-weighted averages (TWAs) for an eight-hour day and forty-hour work week.

Comparing Results to Standards

- Results of the environmental study must be compared with exposure limit standards before control methods can be recommended.

- Exposure limits are the crux of most occupational health codes, regulations, and standards.
- The toxicity of a substance expressed in terms of a threshold limit value (TLV) is an important component in evaluating the presence of a health hazard.
- Typically, if there is no exposure to a harmful agent, it is concluded that its presence does not create a health problem.
- Legislative exposure limits, referred to as permissible exposure limits (PELs), can be found in the Code of Federal Regulations (CFR), Title 29, Part 1910, Occupational and Environmental Health Standards.
- Recommended exposure standards, referred to as threshold limit values, can be found in the *TLVs and BEIs, Threshold Limit Values for Chemical Substances and Physical Agents, Biological Exposure Indices*, published by the American Conference of Governmental Industrial Hygienists (ACGIH).

Exposure Standards

Permissible Exposure Limit (PEL)

- The maximum time-weighted average exposure allowed by OSHA for an eight-hour workday during a normal 40-hour workweek is called the permissible exposure limit (PEL). This is the regulatory synonym for the TLV, which is a recommended exposure limit.

Threshold Limit Value-Time-Weighted Average (TLV-TWA)

- The Threshold Limit Value is based on a time-weighted average (TWA) concentration for a normal eight-hour workday and a 40-hour workweek, to which nearly all workers may be repeatedly exposed, day after day, without adverse effect. It is referred to as the threshold limit value, the TLV, or the TLV-TWA.

Threshold Limit Value-Ceiling (TLV-C)

- The concentration that should not be exceeded during any period of the working exposure is called the Threshold Limit Value-Ceiling (TLV-C).
- In conventional industrial hygiene practice, if instantaneous monitoring is not feasible, then sampling over a 15-minute period (except for those substances to which even short exposures may cause immediate irritation) can assess the TLV-C.

Threshold Limit Value-Short Term Exposure Limits (TLV-STEL)

- The Threshold Limit Value-Short Term Exposure Limit (TLV-STEL) is the concentration to which workers can be exposed continuously for a short period of time without suffering from:
  - Irritation.

- ❑ Chronic or irreversible tissue damage.
- ❑ Narcosis of a sufficient degree to increase the likelihood of accidental injury, impair self-rescue, or materially reduce work efficiency, provided that the daily TLV-TWA is not exceeded.

❑ The TLV-STEL is not a separate, independent exposure limit. It supplements the time-weighted average (TWA) limit where there are recognized acute effects from a substance whose toxic effects are primarily of a chronic nature. STELs are recommended only where toxic effects have been reported from high short-term exposures in either humans or animals.

❑ A TLV-STEL is defined as a 15-minute TWA exposure that should not be exceeded at any time during a workday even if the eight-hour TWA is within the TLV-TWA.

❑ Exposures above the TLV-TWA up to the STEL should not be longer than 15 minutes and should not occur more than four times per day. An averaging period other than 15 minutes may be recommended when this is warranted by observed biological effects.

❑ There should be at least 60 minutes between successive exposures in this range.

UNDERSTANDING THE LIMITATIONS OF EXPOSURE STANDARDS

❑ When comparing sampling results to regulatory or recommended exposure limits, the following must be considered:

- ❑ The dose-response relationship of the agent and the worker.
- ❑ Hypersusceptibility that some workers may experience despite levels within the TLV or PEL.
- ❑ A TLV does not definitely determine a safe versus dangerous condition.
- ❑ Skin absorption potential.
- ❑ The impact of ceiling limits.
- ❑ The biological effects of some agents.
- ❑ Time-weighted-averages for specific substances, unless designated by special categories or ceiling limits, permit limited excursions above the TLV-TWA.
- ❑ The additive or synergistic effects produced in the presence of multiple chemicals.

TLVs and Dose Response

❑ Threshold limit values are based on a dose-response relationship between the agent and the worker. The application of this knowledge to assure that workers are not exposed to concentrations above these threshold values is an important concept in the prevention of occupational diseases.

Hypersusceptibility

- Despite maintaining levels of exposure below a designated TLV, there is no assurance that an individual worker may not show some deleterious effects if hypersensitive to the substance.
- Individuals may be hypersusceptible to some chemicals due to genetic factors, age, personal habits, medication, or previous conditions.

Safe versus Hazardous

- Threshold limit values (TLVs) refer to airborne concentrations of substances and represent conditions under which it is believed that nearly all workers may be repeatedly exposed day after day without adverse effects.
- The TLV of a material is not a fine line between safe and dangerous concentrations. Due to individual variations in susceptibility and the many unknown factors in the working environment, some workers may not be adequately protected from adverse health effects from exposures to certain chemicals at concentrations below the TLV.
- Because of a wide range in individual susceptibility, a small percentage of workers may experience discomfort from substances at or below the threshold limit values. Others may have pre-existing conditions that may be aggravated by exposures below the TLV. For example, workers with asthma may aggravate this medical condition in the presence of sulfur dioxide, even though sulfur dioxide concentrations may be below the TLV.

Biological Effects

- Biologic standards, i.e., the concentration of a specific substance in the urine or blood, represent the body's burden of that agent and may be used as a monitor of the exposure of a worker to a specific agent.
- Biological exposure indices can be found in the *TLVs and BEIs, Threshold Limit Values for Chemical Substances and Physical Agents, Biological Exposure Indices*, published by the American Conference of Governmental Industrial Hygienists (ACGIH).

Additive and Synergistic Effects

- When multiple chemicals or mixtures of toxic substances are present in the absence of other information, their effects are considered additive.
- The combined effect of multiple chemicals or mixtures should be given primary consideration, rather than the effect of any of them individually.

Control of Potential Hazards

CONTROL METHODS

- Reducing or eliminating harmful exposure can control health hazards. Control methods for health hazards in the workplace are classified into three basic categories: engineering controls, administrative controls, and personal protective equipment (PPE).

ENGINEERING CONTROLS

- Engineering controls limit hazards by implementing new design specifications or by applying methods of substitution, isolation, enclosure, ventilation, and work practices.

- Engineering controls are the primary and most desirable method of control, especially for environmental hazards.

Substitution

- Substitution, or replacement of a toxic material with a harmless one, such as using solvent with a lower order of toxicity or flammability, is a very common strategy for minimizing solvent exposure.

Changing a Process

- Changing a process or how a job is performed is a very effective control method. Degreasing an aerospace part by dipping into a solvent tank is preferable to spraying the aerospace part with an atomized solvent.

Isolation

- Isolating employees from a high noise source (either by providing a sound barrier around the noise source or by providing a sound proof booth for workers to be isolated from the noise) is ideal.

Enclosure

- Enclosure is another method of minimizing the escape of air contaminants into the workroom atmosphere.

Work Practices

- Wetting an operation is an effective method for minimizing the source of dust. This can be accomplished by washing down work areas rather than sweeping or dusting off a work area.

Ventilation

- Ventilation is one of the most beneficial methods for minimizing the release of airborne contaminants into the workplace.

- A local exhaust system that traps and removes the air contaminant near the generating source is more effective than general ventilation.

ADMINISTRATIVE CONTROLS

- Administrative controls reduce employee exposures by alternating work schedules or by scheduling reduced work times in contaminant areas. Whenever engineering controls to reduce air contaminants are not feasible, administrative controls may be applied.

- Caution must be used when applying administrative controls, as they have been criticized as a means of spreading exposure to more workers rather than resolving the exposure problem.
- Examples of administrative controls include arranging work schedules and the related duration of exposures so that the employees are minimally exposed to health hazards, and transferring employees who have reached their upper permissible limits to an environment where no further exposure will be experienced.
- Where exposure levels exceed the PEL for one worker in a single day, the job can instead be assigned to two, three, or as many employees as are needed to keep each one's duration of exposure within the PEL.

PERSONAL PROTECTIVE EQUIPMENT

- When effective engineering control methods cannot render a work environment safe, personal protective equipment (PPE) may be required.
- Personal protective equipment (PPE) is specific clothing, gear, or protective devices worn by workers to protect them from a contaminated environment.
- PPE is the least effective method of control and does nothing to reduce or eliminate the presence of the hazard.
- PPE is the least desirable method of control since it relies heavily on human implementation or intervention. If the barrier fails, or if the worker refuses to use the equipment, exposure occurs.

Environmental Factors or Stresses

ENVIRON-MENTAL FACTORS

- The diverse environmental factors that cause illness, impaired health, or significant discomfort in workers can be classified as chemical or physical.

CHEMICAL HAZARDS

- Most occupational health hazards are a result of inhalation of chemical agents in the form of gases, vapors, mists, fumes, and dusts, or by skin contact with these agents. Risks associated with exposure depend on the magnitude and duration of exposure.
- To effectively evaluate and measure a worker's potential exposure, the chemicals in use in any work process must be known. A working knowledge of the physical properties, nomenclature, and effects of exposure are critical to making a proper chemical hazard assessment.

- To learn about chemicals used in a work process, review the raw materials and the nature of the products and by-products manufactured by referring to the products' material safety data sheets (MSDSs).

- The severity of a chemical hazard depends on many factors, including how a chemical is to be used, the type of job performed, duration of exposure, and work patterns.

INHALATION HAZARDS OF CHEMICALS

- When reviewing workplace chemicals for possible inhalation hazards, industrial hygienists must determine whether the chemicals used are soluble or insoluble, asphyxiants, systemic toxins or poisons, irritants, corrosives, teratogens, mutagens, or carcinogens.

Carcinogens

- A carcinogen is a substance or chemical capable of inciting or producing cancer. Benzene is an example of a chemical carcinogen identified as a leukemogenic agent and is notorious for its effects on the blood-forming tissues of the bone marrow.

Solubility

- The solubility of a chemical determines where the chemical deposits become trapped in the respiratory system. The inhalation of highly soluble chemicals, such as sulfur dioxide, may irritate the upper respiratory tract. The inhalation of highly insoluble chemical hazards, such as nitrogen dioxide, may irritate the lower respiratory tract by penetrating the terminal passages of the lungs and the air sacs.

Asphyxiants

- Asphyxiants are classified as simple or chemical. Some chemicals, such as nitrogen gas or Halon®, are considered simple asphyxiants and may dilute the atmospheric oxygen below the level required to sustain the normal blood saturation level. Gases and vapors considered chemical asphyxiants, such as hydrogen cyanide and carbon monoxide, prevent the blood from carrying oxygen to the tissues or interfere with its transfer from blood to tissue.

Systemic Toxins

- Systemic toxins may enter the body through the respiratory system, then enter the bloodstream and produce adverse health effects in a specific organ of the body. Lead is an example of a systemic toxin that can be inhaled and can damage the blood-forming bone marrow and the urinary, reproductive, and nervous systems.

CONTACT HAZARDS OF CHEMICALS

- When reviewing workplace chemicals for possible skin contact hazards, determine whether the chemicals used are irritants or corrosives, or whether they are capable of penetrating the skin by absorption.

- Petroleum-based solvents are examples of skin irritants. Contact may cause the skin to appear white as it defats the skin's oils, but long term damage is minimal.

- Corrosive chemicals may destroy the skin's surface and internal tissues. Hydrofluoric acid is a particularly dangerous corrosive when in direct contact with the skin.

- Some chemicals can directly penetrate the skin, be transported throughout the body, and produce a systemic reaction. Benzene and carbon tetrachloride (CCl_4) are two examples of chemicals with skin absorption potential.

PHYSICAL PROPERTIES OF CHEMICALS

- Some chemicals or products used in the workplace may, under specific circumstances, cause injury to persons or damage to property because of their reactivity, instability, spontaneous decomposition, flammability, or volatility.

Explosives

- A mixture, substance, or compound that is capable of entering into a combustion reaction so rapidly and violently as to produce an explosion.

Corrosives

- Corrosives are capable of destroying living tissue and have a ruinous effect on other substances (especially on combustible materials, which can result in a fire or explosion).

Flammability

- Flammable liquids are liquids with a flash point of 100°F or less.

Oxidizing Agents

- Oxidizers are chemicals that produce oxygen when they decompose. They may create a fire when in contact with a combustible substance, can react violently with water, and can react violently in a fire.

Pressurized Chemicals

- Chemicals stored under pressure, such as compressed gases used for welding, create unique physical hazards. These compressed gases are usually stored in compressed gas cylinders. Extreme caution must be practiced to store these cylinders upright and locked in place.

PHYSICAL STRESSES

- To effectively evaluate a workplace, the physical stresses, as well as the chemical hazards, of the work environment, must be understood.

- Workplace problems associated with noise, temperature extremes, ionizing and non-ionizing radiation are examples of physical stresses or physical hazards.

Noise Hazards

- High noise levels in the workplace are physical stresses that may produce psychological effects by annoying, startling, or disrupting the concentration of a worker.
- High noise levels interfere with human speech, job performance, and safety. Physiological effects of noise include tinnitus, aural pain, and noise-induced hearing loss.

Thermal Stress

- Temperature extremes in the workplace affect the amount of work employees can perform and the method by which they perform their work.
- The body can only operate effectively within a very narrow range of temperature. In high temperature environments, the body must rid itself of excessive heat as quickly as possible to maintain its normal temperature range. In extremely cold environments, the body uses other mechanisms to maintain adequate core body temperatures.

Ionizing Radiation

- Five types of ionizing radiation—alpha, beta, x-ray, gamma, and neutron—may be found or produced in the workplace, subjecting workers to ionizing radiation exposure. These radioactive materials emit energy that can damage living tissue.
- Different types of radiation require different control methods and can cause various external and internal hazards. External hazards come from certain materials that can be hazardous even when located some distance away from the body. Internal hazards enter the body through inhalation, ingestion, or injection.

Non-ionizing Radiation

- Non-ionizing radiation is a form of electromagnetic radiation with varying effects on the body, depending on the wavelength of the radiation involved.
- Non-ionizing radiation includes extremely low frequencies, radio frequency, microwave, ultraviolet light, visible light, and infrared radiation.
- The predominant adverse health effect from non-ionizing radiation is thermal stress, or heating of the body tissues, related to the wavelength, power intensity, and exposure duration. Generally, longer wavelengths produce a greater penetration and more significant temperature rise in tissues than shorter wavelengths.

Ergonomics

- The human body can endure significant discomfort and stress. However, when awkward conditions or motions continue for prolonged periods, they can exceed a worker's physiological limitations.

- Ergonomic considerations include the total physiological and psychological demands of a worker's job, rather than just productivity, health, and safety. Ergonomics includes the biomechanics, workplace design, and behavioral components of a job. Benefits include minimizing physical stress on workers by reducing repetitive motion injuries, reducing injury rates, more efficient operations, lower operating costs, reducing training times, and more effective uses of human resources.

Workplace Design

- Workplace design considers the job and the equipment used by workers, and uses a workplace survey to identify ways to minimize human injury, increase worker efficiency, and decrease human error.

Biomechanics

- Biomechanics deals with the structural elements of the body and the effects of the external and internal forces on various parts of the body. This includes reviewing flexion and extension positions of the hand, elbow joints, and shoulders while performing work. Flexing motions of the joints repeated rapidly over a long period of time results in inflammation of the tendon sheaths, which may cause tendonitis, trigger finger, carpal tunnel syndrome, epicondylitis, and thoracic outlet syndrome, among other illnesses.

Mechanical vibration

- Vibration to the extremities caused by using vibrating power tools, such as jackhammers, reduces circulation in the hands and causes Raynaud's Syndrome, or "white fingers."

Lifting

- When reviewing jobs that require lifting consider task variables, human variables, and environmental variables.

Static Work

- Jobs that require very little movement of the major muscle groups in the body can be as discomforting and fatiguing as highly repetitive jobs. For example, a cramped working posture generates excessive static muscular loading.

BIOLOGICAL HAZARDS

- Where biological hazards are a concern, familiarity with work environments is essential. Examples include:
  - Workers in wood processing facilities exposed to endotoxins.
  - Workers exposed to allergenic fungi growing on timber.
  - Workers associated with birds exposed to *Chlamydia psittaci.*
  - Agricultural workers exposed to anthrax.

Types of Biological Agents

- There are more than 200 biological agents known to produce infections and allergenic, toxic, or carcinogenic reactions in workers.
- These include:
  - Microorganisms such as viruses, bacteria, and fungi capable of producing infection, exposure, and allergies.
  - Allergens and toxins generated from plants that produce dermatitis, rhinitis, or asthma.
  - Protein allergens from vertebrate animals such as dander, hair, and saliva.
  - Arthropod-induced illness such as allergic reaction or inflammation caused by a bite or sting from an insect.
  - Parasite-induced illness such as schistosomiasis caused by the protozoa *schistoma* or roundworm infestations caused by *ascaris*.

2 Particulates: Significance, Sampling, and Analysis

The significance of exposure to particulates depends on critical exposure factors. How the particulates are generated, how they enter the human body, how they impact worker's health, and how to evaluate particulate-laden environments is critical for an industrial hygienist to understand in order to implement effective control measures for particulate exposures.

Types of Particulates

HEALTH HAZARDS OF PARTICULATES

- ❑ Particulates in the work environment can be inhaled. Inhaling too much dust can cause pneumoconiosis.
- ❑ Pneumoconiosis is a lung disease created when inorganic particulates enter the lung.
- ❑ Once in the lung, scarring or fibrosis form around the inorganic particulates, reducing elasticity of the lung and lung functions.
- ❑ Dusts, fumes, fibers, mists, smoke, and aerosols are all particulates.
- ❑ Particulates vary in size and in the manner in which they are created.

Fumes

- ❑ Fumes consist of very small particles, less than a micron (µm) in diameter, usually ranging from 0.001-0.2 µm. Fumes are formed during condensation from a vapor state.

Smoke

- ❑ Smoke is an aerosol (suspended particle in air) formed from the combustion of organic material. Smoke ranges from 0.01-0.5 µm in size.

Dusts

- ❑ A mechanical process such as crushing or grinding generates dusts. Dusts range in size from 0.5-50µm.
- ❑ Common workplace dusts include silica sand from abrasive blasting operations.

Mists

- ❑ Mists are finely divided suspended liquid particles in air formed by condensation from vapor or by some mechanical dispersion process such as atomizing. Mists range in size from 40-400 µm. In comparison, a raindrop equals 100µm.

- Oil mists from machining operations and spray-painting processes are common sources of workplace mists.

Fibers

- Fibers are longer than they are wide and have an aspect ratio greater than 3:1, although defined by the National Institute of Occupational Safety and Health (NIOSH) as having an aspect ratio greater than 5:1. Fibers are sized by their aspect ratio. Aspect ratio = Length/Width (L/W).

- Fibers are generated from crushing, cutting, and mining. Common workplace fibers include asbestos and refractive ceramic fibers.

Critical Exposure Factors

CRITICAL EXPOSURE FACTORS AND THEIR HEALTH IMPACT

- Health problems associated with dust exposure depend upon four critical factors:
  - Type of dust.
  - Length of exposure time.
  - Concentration of airborne dusts in the breathing zone.
  - Size of the dust particles.

- Consider these four factors when determining dust exposure.

- Use measurement techniques to specifically identify the type and size of a particle. Total dust measurement provides very little information about health consequences of the exposure.

Type of Particulate Involved

- Organic dusts:
  - Plant, animal, and/or synthetic origin
  - Produce allergic responses, dermatitis, or systemic toxic effects.

- Inorganic dusts:
  - Metallic and non-metallic.
  - Metallic dusts can produce local dermatitis and sensitization (nickel itch).
  - Non-metallic silica bearing dusts normally are not an acute hazard either to the skin or systematically. Chronic exposure to dust containing crystalline or "free" silica can cause pneumoconiosis.

Length of Exposure

- Pneumoconioses, such as silicosis, asbestosis, and coal worker's pneumoconiosis result, from chronic exposure over several years.

- Exposure to metal fumes, such as zinc and magnesium, may result in metal fume fever after only a few hours. Exposed workers may display flu-like symptoms.

Particulate Concentration

- The actual concentration of dust in the breathing zone is another critical factor.
- Permissible air concentration limits are established as threshold limit values (TLVs) by the American Conference of Governmental Industrial Hygienists (ACGIH) and the permissible exposure limits (PELs) adopted by the Occupational Safety and Health Administration (OSHA).
- Generally, particulates are measured on filters with a measured volume of filtrated air and reported in milligrams per cubic meter (mg/m^3).

Particle Size

- Industrial dusts vary widely in size, with small particles outnumbering larger ones. Therefore, when dust is noticeable in the air around an operation, more microscopic dust particles are also present.
- Only very small particles are respirable or capable of penetrating the lung.
- It is possible to collect respirable sized particles ($<5\mu m$) on a filter, by using a size selective device (cyclone) ahead of a filter at a specific sampling rate.

Particulate Characteristics in the Atmosphere

BEHAVIOR IN THE ATMOSPHERE

- Depending on their size, density, and shape, dust particles settle at various rates, despite the force of gravity. Small particles settle more slowly than larger particles, since small particles are less dense and influenced by Brownian movement.
- Respirable particles released into the air remain airborne for many hours. If dust generation is continuous, the dust level increases along with exposure unless a control method is in place.

Aerodynamic Diameter

- The preferred diameter or size of a particle is considered its equivalent or aerodynamic diameter.
- A particle falls with a steady velocity known as its terminal velocity.
- The terminal velocity is proportional to the specific gravity of the particle, p, and the square of its diameter, d.

- When diameter is expressed in microns, the terminal velocity of a spherical particle falling freely in air is 0.003 pd^2 cm/sec, or 0.006 pd^2 ft/min.

- Settling velocity is the same as terminal velocity.

- It is important to understand aerodynamic diameter when calculating terminal settling rates of particulates.

- The aerodynamic diameter is the diameter of a unit density sphere with the same settling velocity as the particle in question.

- Constants for these calculations include:
  - 1 gm/cm3 = unit density
  - Gravity = 32.2 ft/sec^2 or 98 cm/sec^2.

Fate of Particles

- Stoke's Law applies to the fate of particulates in the atmosphere.

- Stoke's Law is expressed as:

$$V_s = \frac{(\rho - \rho_A)\ gd^2}{18\eta}$$

where:

V_s = terminal settling velocity (cm/sec)
ρ = density of particle (g/cm^3)
ρ_A = density of air (0.00117 g/cm^3)
g = acceleration of gravity (981 cm/sec)
d = diameter of particle (cm)
η = viscosity of air at 25° C

- This equation is applicable to particles between 1 μ and 50 μ, and is greatly influenced by shape (sphere, fiber, etc.).

- Assumes quiescent room environment, which is important when using impingers.

- Reference point is water at 1gm/cm3. As particles get smaller, they become more easily resuspended, usually < 1 μ in size.

Settling Velocity, Simplified

- A simplified formula for settling velocity is expressed as follows:

  $V_s = 0.003\text{cm/sec}(SG)d^2$

  or

  $V_s = 0.006\text{ft/min}(SG)d^2$

where:

V_s = velocity (min or sec)
d = diameter (ft or cm)
SG = specific gravity

Dust Statistics

- Industrial hygiene applies statistical principles to minimize error when sampling and to analyze dust sample characteristics.

- Consider the following statistical principles for particulate analysis:

  - Particulate distribution is generally assumed to be lognormal.

  - These distributions give a straight line when plotted on log-probability paper. The y-axis is the log of the particle diameter and the x-axis is the cumulative percentage of particles with that diameter or less.

  - Using this graph, you can obtain the various percentages of particle size, and calculate the GSD, a dimensionless number.

  - Log normal distribution parameters:

    Geometric mean (GM) = 50% size
    Geometric standard deviation (GSD) = 50%/16% or 84%/50%

- Example: A size analysis on a dust sample indicated that 84% of the particles had a size of 3.8 microns or less and the median particle size was 2.1 microns. What is the GSD of this dust sample?

  $GSD = 3.8/2.1 = 1.81$

  If the 84th percentile was 3.8μ and the 50th percentile was 2.1μ, the GSD for this sample is 1.81.

Behavior in the Respiratory System

SOLID PARTICLES

- Solid particles are stopped in the respiratory system at different levels by various mechanisms.

Mouth and nasal passages

- Impaction and sedimentation remove most particles larger than 10μm. Ragweed pollen (18-25 μm in diameter) in the upper respiratory system can cause hay fever.

Tracheo-bronchial tree

- Impaction, sedimentation, and diffusion remove particles 4 μm and larger.

Alveolar region

- Diffusion removes particles in the 0.5 μm range.

LIQUID PARTICLES

- Liquid particles in the respiratory tract behave as a function of their water solubility and reactivity.
- Mucociliary escalator occurs in the tracheo-bronchiole area to dissolve water-soluble material, then moves to the blood/lymph system and may become systemic.

LUNG CLEARANCE

- Once particles enter the lungs, they may pass to the blood/lymph system. In that system, they are subject to phagocytosis. At the alveolar level, phagocytes containing asbestos will also enter the lymph/blood system.

BIOLOGICAL REACTIONS

- Lung diseases caused by the body's reaction to an accumulation of dust in the lungs include:
  - Fibrosis—scar formation.
  - Bronchitis—overproduction of mucous.
  - Asthma—constriction of bronchial tubes.
  - Cancer—cell mutation.

Systemic Reactions

- Blood absorbs inorganic toxic dusts, lead, and/or mercury.

Metal fume fever

- Inhalation of the metal fumes of zinc, magnesium, and/or copper oxides.

Allergic Reactions

- Skin contact or inhalation of organic dusts such as flour, grains and woods.

Bacterial and Fungus Infections

- Inhalation of active organisms attached to dust particles, wool, or fur dust containing anthrax spores.

Irritation of Nose and Throat

- Inhalation of irritating dusts or mists, such as soluble chromate dusts.

Internal Tissue Damage

- Inhalation of particles that are radioactive, such as radium and its products.

Collection of Particulates

COLLECTION MECHANISMS

- Use the following collection mechanisms when evaluating the atmosphere for particulates.
  - Impaction.
  - Sedimentation.
  - Diffusion.
  - Electrostatic attraction.
  - Direct interception.

Collection Filters for Particulates

- The most common collection device for particulates is the filter. There are several types, including mixed cellulose ester (MCE), glass fiber (GF), Teflon, polyvinyl chloride (PVC), and silver filters.

- Selection depends on the filters' ability to collect material and subsequent laboratory analysis.

- Efficiency is dependent on particle size, shape, specific gravity, and surface area.

- The typical collection device is a closed face filter cassette, 37 mm in diameter, containing a filter supported with a cellulose backup pad.

- Connect the device to a sampling pump and place on a worker's shirt collar.

Commonly Used Particulate Collection Filters

- Mixed cellulose ester (acetate and nitrate) membrane filter:
  - May pick up moisture.
  - Can be dissolved in acid and wet ash.
  - Most common filter used.
  - 0.8 μm pore size, 37 mm diameter.
  - Flow rate: 1.5–2.5 lpm.
  - Unsatisfactory for gravimetric determinations.
  - Used for metals and asbestos.
  - Disadvantage—Tendency to absorb moisture.

- Polyvinyl chloride (PVC) filter:
  - Pore size: 5μm.

- Used primarily for gravimetric determinations on total dust, coal dust, and respirable dust.
- Less pressure drop than mixed cellulose ester filters (MCE).

- Silver filter:
  - Has specialized uses.
  - Does not adsorb materials.
  - Is chemically inert.
  - Benzene soluble material.
  - X-ray diffraction for free silica.

Impingers Used to Collect Particulates

- Impingers are sometimes used for sampling workplace particulates.

- The Greenburg-Smith Impinger operates at a flow rate of 1 cfm, with the nozzle being 2.3 mm diameter, and pressure drop of 3 inches water.

- The standard midget impinger operates at a 1.0 liter per minute (lpm) flow rate and has the following qualities:
  - At 12” water vacuum, the velocity through the jet is 60 m/sec.
  - Particles < 1 μm may remain in air stream.
  - Large particles may break up.

- Some problems that occur may include freezing, foaming, or evaporation.

Electrostatic Precipitators

- Electrostatic precipitators (EP) use an electrical charge to remove particles from the sampled air.

- As the particles pass through the electric field, they become charged and attracted to an oppositely charged electrode.

- Electrostatic precipitators are used in series because collection efficiencies increase with the length of passage through the collector.

- Electrostatic precipitators have low-pressure drop, use low flow rates, and are extremely efficient for smaller particles.

Thermal Precipitators

- Thermal precipitators operate on a thermal gradient. As particles move from a hot atmosphere, they collect on a cold plate.

- Thermal precipitators are very efficient for particles > 0.05μm to 1 or 2μm. The typical sample rate is 7cc/min-0.5 lpm.

Inertial Impactors

- Inertial impactors collect particles by impacting them onto a surface.
- As an air stream moves and changes course, the particles travel straight into a filter, liquid, or, in the case of biological aerosols, an agar plate.
- Efficiency depends on particle size and density, gas velocity, and target size.
- There are two types of inertial impactors—one with a jet against plate and one where the rod moves in air.
- Inertial impactors can be used to determine particle size distribution by starting with a large opening. Gas is then directed to a second jet at a faster velocity that catches the next size particle, etc.
- Cascade impactors generally have three stages; but some have five or six stages. Larger particles collect in the first stages, and smaller particles pass through and collect in the later stages. Particles may bounce off a plate; apply sticky material to prevent this.

Elutriators

- Use an elutriator in front of a sampling train to remove coarse particles. In this process, gravity removes the particles. The smaller particles remain suspended and are collected for laboratory analysis. There are two types: horizontal and vertical.
- Use horizontal elutriators for mineral dusts and vertical elutriators for cotton dusts.

Respirable Sampling

SIZE SELECTIVE TLVS

- The potential hazard for chemical substances present in inhaled air as suspensions of solid particles or droplets depends on particle size as well as mass concentrations because of:
  - The effects of particle size on the deposition site within the respiratory tract.
  - The tendency for many occupational diseases to be associated with material deposited in specific regions of the respiratory tract.
- The ACGIH recommends particle size-selective TLVs for crystalline silica because of the well-established association between silicosis and respirable mass concentrations.

Particle Size-Selective Sampling Criteria

- Particle size–selective TLVs (PSS-TLVs) are expressed in three forms:
  - Inhalable particulate mass TLVs (IPM-TLV) for materials that are hazardous when deposited anywhere in the respiratory tract.
  - Thoracic particulate mass TLVs (TPM-TLV) for materials that are hazardous when deposited anywhere within the lung airways and the gas-exchange region.
  - Respirable particulate mass TLVs (RPM-TLV) for materials that are hazardous when deposited in the gas-exchange region.

Inhalable Particulate Mass (IPM)

- Inhalable particulate mass (IPM) consists of particles that are captured according to the following collection efficiency, regardless of sampler orientation with respect to wind direction:

$$SI(d) = 50\% \text{ x } (1 + e^{-0.06d})$$

for $0 < d <= 100\mu m$

where:

SI(d) = the collection efficiency for particles with aerodynamic diameter d in μm

Particle Aerodynamic Diameter (μm)	IPM (%)
0	100
1	97
2	94
5	87
10	77
20	65
30	58
40	54.5
50	52.5
100	50

Thoracic Particulate Mass

- Thoracic particulate mass (TPM) consists of particles that are captured according to the following collection efficiency:

$$ST(d) = SI(d)\,[1\text{-}F(x)]$$

where:

$$x = \frac{\ln (d/\Gamma)}{\ln (\Sigma)}$$

$\Gamma = 11.64$

$\Sigma = 1.5$

F(x) = the cumulative probability function of a standardized normal variable, x

Particle Aerodynamic Diameter (μm)	TPM (%)
0	100
2	94
4	89
6	80.5
8	67
10	50
12	35
14	23
16	15
18	9.5
20	6
25	2

Respirable Particulate Mass

- Respirable particulate mass (RPM) consists of those particles that are captured according to the following collection efficiency:

$$SR(d) = SI(d)\,[1 - F(x)]$$

where:

F(x) has the same meaning as above with $\Gamma = 4.25$ μm and $\Sigma = 1.5$

Particle Aerodynamic Diameter (μm)	RPM (%)
0	100
1	97
2	91
3	74
4	50
5	30
6	17
7	9
9	5
10	1

Respirable Dust Sampling

- The major difference here is the increase in the median cut point for a respirable particulate matter sampler from 3.5 μm to 4.0 μm.

- No change in sampling has been recommended by ACGIH. The measurement of respirable particulates using a 10-mm nylon cyclone at a flow rate of 1.7 liter per minute remains.

- Two analyses of available data indicate that the flow rate of 1.7 liters per minute allows the 10-mm nylon cyclone to approximate the particulate matter concentration that would be measured by an ideal respirable particulate sampler.

Cyclone Collection Device

- A cyclone is used to collect particles of respirable size.
- Respirable particles are those that are retained in the lung and are generally considered to be of an aerodynamic size below 10 µm.
- Air is drawn into a cyclone tangentially through a small opening. Then:
  - The air moves through the cyclone in centrifugal motion and forces the larger particles to the outside edge of the air stream, where they fall to the bottom of the cyclone.
  - The respirable particles stay in the center of the air stream and are drawn upward onto a pre-weighed filter.
- After sampling, the filter is again weighed to determine how much material has been collected. This is termed gravimetric analysis.

Stack Sampling

- Sampling for air pollution particulates from exhaust stacks requires isokinetic sampling.
- Isokinetic condition means that the gas velocity entering the sampling probe is equal to that of the sample stream.
- If the air stream is faster than the air entering the collection device, the sample will yield a bias to bigger particles. The opposite is true when the air stream is slower than the air entering the collection device, which will yield a bias to smaller particles.

Particulate Air Sample Analysis

ANALYTICAL PROCEDURES

- There are several methods for analysis of particulates:
  - Optical microscopy: Dust, asbestos, fibers.
  - Gravimetric: Total dust, free silica, coal dust.
  - Wet chemical: Free silica, lead, isocyanates.
  - Instrumental: Atomic absorption (metals), atomic spectroscopy.
  - Direct-reading field instruments: Aerosol photometers, beta absorption instruments, Piezo-electric instruments.

Asbestos Sampling

- Collect asbestos samples on a 25-mm filter and cassette with an open-face 50-mm conductive extension cowl.
- Using an electrically conductive extension cowl reduces the number of asbestos fibers attracted to the sides of the cassette by static electricity.

- Collect asbestos on a 0.8 µm mixed cellulose ester membrane filter at 1-2.5 lpm (EPA, 16 lpm).

- Cut a 1/8 wedge from the filter at the time of analysis and mount on a slide with 1:1 dimethyl phthalate and diethyl oxalate.

- Use phase contrast microscopy (PCM) to count all fibers longer than 5µm and having an aspect ratio greater than or equal to 3:1 (NIOSH = 5:1).

- Count as many fields as necessary to total 100 fibers (at least 20 but no more than 100).

Fiber Counting Using a Graticule

- Calculation of an asbestos sample result is outlined below using the graticule fiber counting method. A sample was taken with a 25mm MCE filter at 2 lpm for 180 minutes and analyzed using Method 7400. There were 15 fibers per field on the filter. The blank filters came with 0.02 fibers per field. The usable filter area was 385 mm. What was the air concentration?

Method 7400: Walton-Beckett graticule: 100µm dia.

Convert µ to mm - $100\mu m \times \frac{1\ mm}{1000\ \mu} = 0.10\ mm$

$$A = \pi r^2$$

$$= \pi \left(\frac{0.10\ mm}{2}\right)^2$$

$$= 0.00785\ mm^2$$

$$\frac{15\ fibers}{field} \times \frac{1\ field}{0.00785 mm^2} \times \frac{385\ mm^2}{filter} \times \frac{1\ filter}{\{2\ l/m \times 180\ min/l\}} \times \frac{1\ l}{1000\ cc}$$

$$= 2.04\ fibers/cc$$

Formula for Fiber Counting

- An alternative formula may be used for calculating an asbestos sampling result, expressed as:

$$C = \frac{(C_s - C_b) \times A_s}{A_f \times T \times R \times 1000\ cc/l}$$

where:

C = fiber concentration
C_s = average measured fiber count per field
C_b = average blank fiber count per field
A_s = area of filter (mm^2)
A_f = area of field (mm^2)
T = sampling time
R = sampling rate (liters per minute)

- Using the information provided in the previous block, the alternative formula yields the same result.

$$C = \frac{(15 - 0.02) \times 385}{0.00785mm^2 \times 2.0 \times 180 \times 1000} = 2.04 \text{ fibers/cc}$$

Particulates of Health Significance

SILICA

- Silicon dioxide usually refers to:
  - Amorphous silica (non-crystalline).
  - Crystallized silica such as sand (quartz).
  - Silicates such as clay (aluminum silicate).

- Only the crystalline (free silica) material found in quartz, tridymite, cristobalite, and a few other non-silicate materials causes silicosis. Silicosis is a lung disease caused by the inhalation of free silica. The risk of silicosis is found in glass manufacturing, mining of quartz rock, and foundries.

- Crystallized silicon dioxide (SiO_2) most commonly occurs in sand. The percentage of crystalline SiO_2 in a dust mixture is the usual basis for evaluating its inhalation hazards.

- Inhaled free silica particles have a toxic effect on macrophage cells (white blood cells). Macrophage cells attempt to engulf and remove foreign material from the lungs. The macrophages cannot entirely engulf the particles and ultimately, fibrosis or scarring occurs around the particle. This silicotic nodule is not as elastic as the normal lung tissue and prohibits the ready passage of oxygen and carbon dioxide.

ASBESTOS

- Asbestos is a general term given to several naturally occurring, fibrous, hydrated mineral silicates that differ in chemical composition.

- Asbestos fibers have high tensile strength, flexibility, heat and chemical resistance, and good friction properties.

- Asbestos can be woven, spun, and pressed into cement, plastic, asphalt, tiles, brake drums, etc.

- Forms of asbestos:
  - Chrysotile (white asbestos) is a fibrous form and the most common variety.
  - Crocidolite (blue asbestos) is also fibrous but less common.
  - Amosite is easily broken down into yellow-brown harsh fibers.
  - Anthophyllite is fragile and brownish.
  - Temolite is used in the production of talc.
  - Actinolite is rarely used in industry.

- Asbestos causes three major diseases:
  - Asbestosis: a diffuse, interstitial nonmalignant scarring of the lungs.
  - Bronchogenic carcinoma: a malignancy of the lining of the lung's air passages.
  - Mesothelioma: a diffuse malignancy of the lining of the chest cavity or the lining of the abdomen.

- Asbestos is manifested by its characteristic appearance on x-rays, by restrictive pulmonary functions, or by clinical signs.

Asbestos Regulations

- The Asbestos Hazard Emergency Response Act (AHERA) has established standards for:
  - Training and accreditation of asbestos professionals.
  - Protocols for building inspections for asbestos-containing materials.
  - Requirements for response actions to asbestos hazards.
  - Clearance tests to determine completion of asbestos abatement activities.

- The subsequent Asbestos Schools Hazard Abatement Reauthorization Act (ASHARA) and Model Accreditation Plan increased training requirements and extended many of the provisions to public and commercial buildings.

- OSHA's General Industry Standard applies to all occupational exposures to asbestos except construction and shipbuilding. Workers affected include those involved in manufacturing asbestos, those engaged in brake and clutch repair, and those involved in floor maintenance where the floors contain asbestos.

Asbestos Regulation Requirements

- Asbestos regulations require comprehensive work practices.
- Asbestos exposure must be maintained to 0.1 fiber/cc in an eight-hour TWA or 1.0 fiber/cc of air for 30-minute short-term excursion limit.
- Monitoring must be performed consistent with the regulations.
- Monitoring must be performed by a competent person, meaning one capable of identifying asbestos hazards in the workplace and selecting the appropriate control strategy for asbestos exposure, and who has completed specialized training to perform these functions.
- Medical surveillance must be provided for applicable exposed workers.
- Personal protection must be provided applicable to the hazard.
- Engineering controls must be in place for the various classes of asbestos exposures.
- Work practices must be established that prohibit:
  - Sanding.
  - High-speed abrasive sawing without high efficiency particulate filtered exhaust ventilation.
  - Using compressed air to cleanup debris.
  - Dry sweeping, shoveling, or other dry cleanup.

COAL DUST

- Black lung is a term identifying all lung diseases associated with chronic overexposure to coal dust. These diseases include chronic bronchitis, silicosis, and coal worker's pneumoconiosis (CWP).
- Exposure to coal dust is regulated by the Mine Safety and Health Administration (MSHA) if exposure occurs in mines, and by OSHA if exposure occurs in other operations.

BERYLLIUM

- Beryllium intoxication is a severe systemic disease that results from the inhalation of dust or fumes from metallic beryllium, beryllium oxide, or soluble beryllium compounds.
- The acute form of beryllium intoxication is chemical pneumonitis with chronic coughing, pain, breathing difficulty, cyanosis, and weight loss.

- Chronic beryllium disease *(berylliosis)* results from inhalation of beryllium particulates. It is characterized by granulomas in the lungs, skin, and other organs. Symptoms include cough, chest pain, and general weakness. Pulmonary dysfunction and systemic effects include heart enlargement.

WELDING FUMES

- The composition and quantity of welding fumes depends on the alloy being welded, the process, and the electrodes used.
- Metal fume fever is an acute condition of short duration caused by brief high exposure to the newly generated fumes of metals such as zinc and magnesium.
- This disease may occur from the inhalation of extremely fine particles that have been freshly formed as fumes.
- Symptoms appear four to twelve hours after exposure and include fever and shaking chills.

Control of Particulates

TYPES OF CONTROLS

- The three most common methods of control for particulates include engineering controls, administrative controls, and personal protective equipment.

Engineering Controls

- The best method for controlling particulates is to enclose dust-generating processes, include exhaust ventilation, and maintain a negative pressure.
- Local exhaust ventilation can be used at points of dust production, and is best combined with a hooded enclosure. A good example is the hooded ventilation system used for buffing and grinding operations, which allows the operator to stand above and outside of the hood.
- If sources of dust are numerous and evenly distributed, general ventilation may be necessary. However, this makes it costly to heat or cool make-up air.
- Moisture control is very effective as an engineering control. Wet drilling and grinding are effective means of preventing dust generation.

Administrative Controls

- Wet methods of housekeeping and wet cleanup methods minimize dust re-generation.

- ❑ Prohibiting the use of compressed air or dry sweeping to clean work surfaces is an excellent administrative control.

- ❑ High Efficiency Particulate Air (HEPA) vacuums are recommended for specific dust hazards, such as lead or asbestos.

- ❑ Prohibiting eating, drinking, or smoking around dusty processes is an effective control.

Personal Protective Equipment

- ❑ Respirators are effective personal protective devices for preventing the inhalation of hazardous dusts.

- ❑ Protective clothing can be worn to prevent hazardous dusts from getting into work clothes and boots and being subsequently carried home by workers.

3 Air Sampling Calculations and Gas Laws

Calculations are critical in the field of industrial hygiene. It is necessary to calculate sample volumes when evaluating a work environment, including sample times, TLVs, air concentrations from vapor pressures, and determining the additive effects of chemicals when multiple agents are used in the work environment. It is sometimes necessary to calculate volume changes as temperatures and pressures change in the work environment.

Air Sampling Calculations

CALCULATING SAMPLE VOLUMES

- Determining how much to sample or collect on a filter or in other media is necessary prior to sampling.

- The formula for calculating minimum sample volume in liters is outlined below:

$$SV = \frac{AS(\mu g)}{PEL\ (mg/m^3)\ \times\ F\ (decimal)}$$

where:

SV = minimum sampling volume (liters)
PEL = permissible exposure limit (mg/m^3)
AS = analytical sensitivity (μg)
F = desired fraction of TLV/PEL (decimal)

Note: The fraction is what you want to see, such as 1/10 below the standard. Please note the units do not appear to cancel out properly; however the formula is correct.

- Using dimensional analysis, the formula is expressed as:

$$SV = \frac{\mu g}{mg/m^3} \times F \times \frac{1\ m^3}{1000L} \times \frac{1\ mg}{1000\ \mu g}$$

Example: Sample Volume Calculation

- What sample volume is needed to "see" CS_2 at a concentration of 1/10th the TLV, given an analytical sensitivity of 10μg?

$$\text{Sampling Volume (SV)} = \frac{\text{Analytical Sensitivity } (\mu g)}{\text{PEL } (mg/m^3) \times \text{Fraction}}$$

Convert ppm to mg/m³ first:

$$mg/m^3 = \frac{MW}{24.45 \times ppm}$$

$$= \frac{76}{24.45 \times 10}$$

$$= 31$$

Therefore,

$$= \frac{10\mu g}{31\ mg/m^3 \times 1/10} \times \left[\frac{1\ mg}{1000\mu g} \times \frac{1000\ l}{1\ m^3}\right] = 3.2\ l$$

Calculating Sample Time

- The minimum sampling time is usually expressed in liters per minute.

$$\text{Minimum sample time (L/min.)} = \frac{SV}{SR}$$

where:

SV = minimum sampling volume (liters)
SR = sampling rate (minutes)

TLV-TWA

- The following formula is used to calculate the threshold limit value (TLV).

$$TLV = \frac{T_1C_1 + T_2C_2 + \ldots + T_nC_n}{T_1 + T_2 + \ldots + T_n}$$

where:

T = time
C = concentration

Note: The answer is a number.

Example: TWA Calculations

- Given: two hours @ 60 ppm, one hour @ 30, four hours @ 100, calculate the TWA.

$$TWA = \frac{T_1C_1 + T_2C_2 + ... + T_nC_n}{T_1 + T_2 + ... + T_n}$$

$$= \frac{(2\ hrs)(60\ ppm) + (1\ hr)(30\ ppm) + (4\ hrs)(100\ ppm)}{2\ hrs + 1\ hr + 4\ hrs}$$

$$= 78.6\ ppm$$

TLV-Mixture Additive Effects

- Use the following formula when calculating the TLV of a mixture with additive or synergistic adverse health effects.

$$TLV = \frac{C_1}{TLV_1} + \frac{C_2}{TLV_2} + \frac{C_3}{TLV_3} + ... + \frac{C_n}{TLV_n}$$

where:

C = Concentration of each component
T = TLV of each component

- The answer is a ratio. If TLV $\leq$ 1, it has not been exceeded. If TLV > 1, it has been exceeded. Unity is equal to 1.

  Note: If effects are not similar, look at individual TLVs.

Example: TLV for Mixtures (Additive or Similar Effects)

- Given: 400 ppm acetone (TLV = 750 ppm)
  150 ppm sec-butyl acetate (TLV = 200 ppm)
  100 ppm methyl ethyl ketone (TLV = 200 ppm)

- What is the TLV for this exposure to a mixture of solvents?

$$TLV_{mixture} = \frac{400\ ppm}{750\ ppm} + \frac{150\ ppm}{200\ ppm} + \frac{100\ ppm}{200\ ppm}$$

$$TLV_{mixture} = 1.78$$

1.78 > 1.0, thus TLV for mixture is exceeded.

TLV-Liquid Mixture

- Use the following formula when a liquid chemical mixture is being collected in the atmosphere.

TLV-Liquid Mixture—Atmospheric composition same as liquid.

$$\text{TLV (mixture)} = \frac{1}{\frac{f_a}{TLV_a} + \frac{f_b}{TLV_b} + \frac{f_c}{TLV_c} + \ldots + \frac{f_n}{TLV_n}}$$

where:

f_n = weight fraction of material in liquid
TLV_n = TLV in mg/m^3

Note: Assume evaporation rates must be the same. Formula does not apply if vapor pressures are very different.

Example: TLV-Liquid Mixture

- Given: Methylene chloride 50% and methyl chloroform 50%; Methylene chloride TLV = 174 mg/m^3; Methylene chloroform TLV = 1910 mg/m^3.

$$TLV = \frac{1}{\frac{0.5}{174} + \frac{0.5}{1910}}$$

$$TLV = \frac{1}{0.29 + 0.00026} = 319 \text{ mg/m}^3$$

Air Concentrations from Vapor Pressure

- Use the following formula when calculating air concentrations from vapor pressure data.

$$ppm = \frac{P_v}{P_{at}} \times 10^6$$

where:

P_v = Partial vapor pressure of contaminant/solvent
P_{at} = vapor pressure of atmosphere

Example: Air Concentration from Vapor Pressure

- How much agent X is in the atmosphere in parts per million if its vapor pressure is 33 mm Hg at normal temperature and pressure (NTP)?

$$X \text{ in ppm} = \frac{33 \text{ mm Hg}}{760 \text{ mm Hg}} \times 10^6$$

$$X \text{ in ppm} = 43{,}421$$

Silica—TLV

- The TLV of a free silica or crystalline silica sample can be found in the TLV booklet published by ACGIH.

TLV (mg/m3) = 0.05 for cristobalite
= 0.1 for quartz
= 0.05 for tridymite
= 0.1 for tripoli (of contained respirable quartz)

Note: TLV is higher than OSHA PEL or NIOSH REL

- Inhalable (total) particulate dust for matter containing no asbestos and <1% crystalline silica TLV (mg/m^3) = 10.0

Silica—PEL

- Calculate OSHA's permissible exposure limits (PELs) by using the following formulae.
- Quartz (respirable):

$$\text{PEL (mg/m3)} = \frac{10\ mg/m^3}{\%\ SiO_2 + 2}$$

- Quartz (total dust):

$$\text{PEL (mg/m3)} = \frac{30\ mg/m^3}{\%\ SiO_2 + 2}$$

- Cristobalite: Use 1/2 the value calculated from the count or mass formulae for quartz.
- Tridymite: Use 1/2 the value calculated from the formulae for quartz.

Silica—PEL for Mixtures

- Calculate the OSHA PEL for mixtures of silicates using the following formula:

$$\text{PEL (mixture)} = \frac{10\ mg/m^3}{\%\ Q + 2(\%C) + 2(\%T) + 2}$$

where:

Q = quartz
C = cristobalite
T = tridymite

Silica—PEL for Mixtures

- Calculate the OSHA PEL for mixtures of silicates using the following formula:

$$\text{PEL (mixture)} = \frac{10\ mg/m^3}{\%\,Q + 2(\%C) + 2(\%T) + 2}$$

where:

Q = quartz
C = cristobalite
T = tridymite

Parts Per Million Calculations

- Calculating the volume of a gas or vapor usually expressed as parts per million (ppm) can be performed using the gas volume in cubic feet or liters.

$$ppm = \frac{V_{gas}}{V_{air}} \times 10^6$$

where:

ppm = concentration in parts per million
V_{gas} = volume of gas (ft^3 or l)
V_{air} = volume of air (ft^3 or l)

Converting ppm to mg/m³

- Converting concentrations of a gas or vapor in part per million (ppm) to milligrams per cubic meter (mg/m3) is performed as follows:

$$mg/m^3 = \frac{\text{Molecular weight}}{24.45} \times ppm$$

`45 (at 25° C and 760 mm Hg)

Converting mg/m³ to ppm

- When converting milligrams per cubic meter to parts per million, the following formula is used.

$$ppm = \frac{24.45}{\text{Molecular weight}} \times mg/m^3$$

Kinetic Molecular Theory and the Gas Laws

IMPORTANCE OF GAS LAWS IN INDUSTRIAL HYGIENE

- Gases and vapors behave similarly.
- They both follow the ideal gas laws in that their volume is affected by changes in temperature and pressure.

Gas Law Theories

- The fact that all gases behave similarly with a change in temperature and pressure led to the Kinetic-Molecular Theory.
    - All matter is composed of discrete particles called molecules.
    - The molecules are relatively far apart in gases.
    - The molecules of all substances are in continuous rapid motion.
    - Collisions between molecules are perfectly elastic. When molecules collide with one another or with the walls of the container, they rebound without any loss in energy. This assumption is necessary to account for the fact that the pressure exerted by a gas does not vary after a long period of time. The molecules must move perpetually and rebound from collisions with no loss in energy to account for these facts.
    - The average kinetic energy of molecules is directly proportional to the Kelvin temperature.

Experimental Facts Based on These Assumptions

- Based on the theories above, the following are true:
    - Gases exert pressure.
    - Gases are highly compressible.
    - Gases diffuse very easily.
    - Pressure exerted by a gas increases with the temperature if the volume remains constant.
    - Gases expand when heated (if the pressure remains constant).

Units of Measurement

- We usually want to know the pressure, temperature, and volume of gases. The most common method for expressing pressure is millimeters of mercury (mm Hg).
- Temperature is measured in degrees Celsius or Kelvin and volume in cubic centimeters (cc), millimeters (ml), and liters (l).
- A barometer measures pressure. A glass tube roughly one yard long is filled with mercury and the open end is inverted into a dish of mercury. The mercury falls in the tube, leaving a vacuum above it, until the pressure exerted by the air just balances the mercury column in the tube. With this method, the atmospheric pressure is expressed as the measurement of the height of the column in millimeters of mercury (mm Hg).

- There must be a common standard for measurement before comparing gases under different conditions of temperature and pressure.
- Standard pressure is 760 torr (760 mm of mercury at sea or atmosphere level).
- Standard temperature is 0° C or 273° K (Kelvin = 273 + x° C).

The Gas Laws

BOYLE'S LAW

- At constant temperature, the volume occupied by a given mass or gas is inversely proportional to the pressure, (as the pressure goes up, the volume of a gas goes down and as the pressure goes down, the volume goes up).
- If the pressure is decreased to half the original pressure, the new volume is double the old volume.
- This is mathematically expressed as follows:

$$P_1V_1 = P_2V_2 \quad \text{or} \quad \frac{V_1}{V_2} = \frac{P_2}{P_1}$$

where:

P_1 and V_1 are the initial pressure and volume
P_2 and V_2 are the final pressure and volume

CHARLES' LAW

- At constant pressure, the volume of a gas is directly proportional to its Kelvin, or absolute, temperature.
- The term Kelvin is very important in this definition and should always be included. Charles, a French Scientist, found that the volume of a gas at 0° C increased by 1/273 for each degree rise in its temperature and decreased 1/273 for each degree that its temperature fell.
- For example, if a gas is cooled 1° C, its volume decreases by 1/273; if it is cooled 273° C, its volume decreases and becomes zero. However, a gas will liquefy before it reaches this temperature, which Charles called "absolute zero."

- At a constant pressure:

$$\frac{V_1}{V_2} = \frac{T_1}{T_2}$$

where:

V_1 and V_2 are the initial and final volumes
T_1 and T_2 are the initial and final Kelvin temperature

COMBINATION OF BOYLE'S AND CHARLES' LAWS

- The above laws can be used to derive the third general gas law.

$$[PV/T]_1 = [PV/T]_2$$

or

$$\frac{P_1V_1}{T_1} = \frac{P_2V_2}{T_2} \quad \text{or} \quad \frac{V_1}{V_2} = \frac{T_1}{T_2} \times \frac{P_2}{P_1}$$

- This combination is useful in calculating volumes to STP as is required in some industrial hygiene situations.

OTHER GAS LAWS OF GENERAL INTEREST TO INDUSTRIAL HYGIENE

- The density of a gas is related to the volume, temperature, and pressure. As the volume increases, the mass per unit volume (density) decreases proportionally. Hence, density varies inversely with volume.

- Secondly, density varies directly with pressure and inversely with absolute temperature.

- This is mathematically expressed as:

$$d_2 = d_1 \times \frac{T_1}{T_2} \times \frac{P_2}{P_1}$$

Dalton's Law of Partial Pressure

- The total pressure of a gaseous mixture is equal to the sum of the partial pressures of the components.

- The partial pressure of each component gas is equal to the pressure it would exert if it occupied the entire volume (most accurate for the "ideal" gases and pressures of a few atmospheres).

- In a mixture of gases, each gas exerts a pressure independent of others and proportional to the relative amount by volume of that gas in the mixture.

- Thus:

  $$P_1 + P_2 + P_3 + \dots P_n = P_{total}$$

Avogadro's Hypothesis—Gram Molecular Volumes

- Avogadro stated that equal volumes of all gases under the same conditions of temperature and pressure contain the same number of molecules.
- From this theory, it follows that the weight of any volume of gas may be compared to any other equal volume of gas. Hence, if one gas weighs twice that of another gas, then its molecular weight (or gram-molecular volume or molar volume) must be twice that of the lighter gas.

Example: Avogadro's Hypothesis

- If oxygen is used as an example to demonstrate Avogadro's hypothesis, the following applies:
  - One molecule of oxygen contains two atoms (O_2) with a molecular weight of 32. Hence, one mole is 32.0 grams.
  - The density of oxygen is 1.429 grams per liter.
  - Thus, the volume occupied by one mole of oxygen is 32g/1.429 grams per liter.
  - Performing the calculation yields 22.4 liters at STP. The significance of this is that with Avogadro's hypothesis, 22.4 liters of any gas will contain the same number of molecules (6.02×10^{23} molecules).

Ideal Gas Law

- Avogadro's hypothesis described above can be expressed as the Ideal Gas Law.

  $$PV = nRT$$

where:

n = number of moles of gas (or mass in grams divided by the molecular weight)
R = molar gas constant
P = pressure
V = volume
T = temperature, always in Kelvin

- The above relationship explains why one-gram molecular weight of gas occupies 22.4 liters of space at standard temperature and pressure (STP). This number represents the number of molecules in one mole (GMW) of any compound.

Sample Calculations and Conversions

Boyle's Law

1. If 400 ml of oxygen is collected at a pressure of 780 mm Hg, what volume will the gas occupy if the pressure is changed to 740 mm Hg?

$$\text{New volume } (V_2) = 400 \text{ ml} \times \frac{780 \text{ mm Hg}}{740 \text{ mm Hg}} = 422 \text{ ml}$$

2. What is the volume of hydrogen at a pressure of 80 cm Hg if 200 ml of the hydrogen was collected at a pressure of 76 cm Hg?

$$\text{New volume } (V_2) = 200 \text{ ml} \times \frac{76 \text{ cm Hg}}{80 \text{ cm Hg}} = 190 \text{ ml}$$

3. Calculate the pressure of a gas that occupies a volume of 100 ml, if it occupies a volume of 200 ml at a pressure of 70 cm Hg.

$$\text{New pressure } (P_2) = 70 \text{ cm Hg} \times \frac{200 \text{ ml}}{100 \text{ ml}} = 140 \text{ cm Hg}$$

Charles' Law

4. What volume will an amount of nitrogen occupy at 27° C if the gas occupies a volume of 400 ml at a temperature of 0° C? Assume that the pressure remains constant.

$$^{\circ}\text{K} = 273^{\circ} + {}^{\circ}\text{C}$$

$$\text{New volume } (V_2) = 400 \text{ ml} \times \frac{300^{\circ} \text{ K}}{273^{\circ} \text{ K}} = 440 \text{ ml}$$

5. What is the volume of a gas at -20° C if the gas occupied 50.0 ml at a temperature of 0° C?

$$\text{New volume } (V_2) = 50.0 \text{ ml} \times \frac{253^{\circ} \text{ K}}{273^{\circ} \text{ K}} = 46.4 \text{ ml}$$

6. If a gas occupies a volume of 700 ml at 10° C, at what temperature will it occupy a volume of 1000 ml if the pressure remains constant?

$$\text{New absolute temperature } (T_2) = 283^{\circ} \text{ K} \times \frac{1000 \text{ ml}}{700 \text{ ml}} = 404^{\circ} \text{ K}$$

Boyle's Law and Charles' Law Combined

7. Calculate the volume of a gas at STP if 500 ml of the gas is collected at 27° C and 75 cm Hg.

$$\text{New volume } (V_2) = 500 \text{ ml} \times \frac{75 \text{ cm Hg}}{76 \text{ cm Hg}} \times \frac{273^\circ\text{K}}{300^\circ\text{K}} = 449 \text{ ml}$$

8. If a gas occupies a volume of 100 ml at a pressure of 76 cm Hg and 27° C, what volume will the gas occupy at 800 mm Hg and 50° C?

$$\text{New volume } (V_2) = 100 \text{ ml} \times \frac{760 \text{ mm Hg}}{800 \text{ mm Hg}} \times \frac{323^\circ\text{K}}{300^\circ\text{K}} = 102 \text{ ml}$$

9. If 400 ml of oxygen is collected at 20°C, and the atmospheric pressure is 735.0 mm Hg, what is the volume of the dry oxygen at STP?

$$\text{New volume of dry gas} = 400 \text{ ml} \times \frac{735.0 \text{ mm Hg}}{760 \text{ mm Hg}} \times \frac{273^\circ\text{K}}{293^\circ\text{K}} = 360 \text{ ml}$$

Gram Molecular Volumes
Note: 1 mole = 22.4 liters at STP
1 mole = 24.45 liters at NTP.
Pay attention to the conditions. Many of these questions are geared to standard conditions (STP).

10. 2.50 g of a gas occupies 500 ml at STP. What is the molecular weight (MW) of the gas?

The problem is to find the weight of 22.4 liters, since this is the volume at STP occupied by one mole of the gas. We have 2.50 g gas per 500 ml at STP or 2.50 g/500 ml. Changing ml to L we have:

$$\frac{2.50 \text{ g}}{500 \text{ ml}} \times \frac{1000 \text{ ml}}{1 \text{ L}} = 5.00 \text{ g/L}$$

Then, changing liters of gas to moles (one mole of any gas at STP occupies 22.4 l),

$$\frac{5.00 \text{ g}}{1 \text{ L}} \times \frac{22.4 \text{ L}}{1 \text{ mole}} = 112 \text{ g/mole}$$

Since 112 g of the gas at STP occupy 22.4 L, the molecular weight is 112. This may all be combined into one step, or

$$\frac{2.50 \text{ g}}{500 \text{ ml}} \times \frac{1000 \text{ ml}}{1 \text{ L}} \times \frac{22.4 \text{ L}}{1 \text{ mole}} = 112 \text{ g/mole}$$

11. 2.40 g of a gas occupy 250 ml at 20° C and 740 torr. What is the gram molecular weight of the gas?

$$250 \text{ ml} \times \frac{273^\circ \text{ K}}{293^\circ \text{ K}} \times \frac{740 \text{ torr}}{760 \text{ torr}} = 227 \text{ ml, then}$$

$$\frac{2.40 \text{ g}}{227 \text{ ml}} \times \frac{1000 \text{ ml}}{1 \text{ L}} \times \frac{22.4 \text{ L}}{1 \text{ mole}} = 237 \text{g/mole}$$

12. If 7.50 g of a gas occupy 612 ml at 30°C, and 745 torr, what is the MW of the gas?

$$612 \text{ ml} \times \frac{273^\circ \text{ K}}{303^\circ \text{ K}} \times \frac{745 \text{ torr}}{760 \text{ torr}} = 540 \text{ ml, then}$$

$$\frac{7.50 \text{ g}}{540 \text{ ml}} \times \frac{1000 \text{ ml}}{1 \text{ L}} \times \frac{22.4 \text{ L}}{1 \text{ mole}} = 311 \text{g/mole}$$

4 Air Sampling for Gases and Vapors

To effectively evaluate a potentially hazardous workplace, an industrial hygienist must obtain objective and quantitative data. This usually requires some form of air sampling. The type of sampling performed depends on the airborne contaminant in question.

Personal and Area Sampling

TYPES OF AIR SAMPLING

- The type of air sampling performed depends on:
  - Purpose of sampling.
  - Equipment available.
  - Environmental conditions.
  - Nature of the toxic contaminant.

Representative Sampling

- Every effort must be made to obtain samples that represent the worker's exposure.

- To determine what constitutes a representative sample, decide:
  - Where to sample.
  - Whom to sample.
  - How long to sample.
  - How many samples to take.
  - When to sample.

Personal Sampling

- Samples representative of worker exposure must be collected if the purpose of sampling is to:
  - Define a potential hazard.
  - Check compliance with specific regulations.
  - Determine a worker's exposure and daily time-weighted-average.
  - Obtain data for control purposes.

- Personal air sampling is the preferred method of evaluating worker exposure to airborne contaminants.
  - The worker wears a sampling device that collects an air sample.
  - Place the collection device as close as possible to the breathing zone of the worker so the information collected approximates the concentration inhaled. The breathing zone is defined as a hemisphere in front of the shoulders with a radius of six to nine inches.

Area Sampling

- ❑ Area monitoring is very effective for confined spaces or sewer lines.
- ❑ Use area air samples to evaluate background concentrations, locate sources of exposure, and evaluate the effectiveness of control measures.
- ❑ Area samples taken at a fixed location in the workroom must be appropriate for the specific need.
- ❑ Use area samples when an estimation of worker exposure is not necessary.
- ❑ Use area samples to identify high exposure areas, indicate flammable or explosive concentrations, or determine if an area should be isolated or restricted.

Integrated Versus Grab Sampling

INTEGRATED SAMPLING

- ❑ Integrated sampling, the preferred method for determining time-weighted average (TWA) exposures, consists of one or a series of samples taken for the full or partial duration of the time averaging period.
- ❑ Integrated air sampling involves the extraction of a gas or vapor from a sample air stream.
- ❑ Pass a known volume of air through a collection media to remove the contaminant from the sample air stream and deposit the contaminant in or on a medium. Laboratory analysis follows.
- ❑ Use integrated sampling to estimate an eight-hour or 15-minute exposure to a particular substance. Collect one or more personal air samples for the duration of a particular task or work shift.
- ❑ Use integrated samples to combine all of the various concentrations to which the worker has been exposed during the sampling period. Sometimes these concentrations are based on grab samples collected throughout the work shift.

GRAB SAMPLING

- ❑ Grab sampling, or instantaneous sampling, is the collection of an air-sample over a short period of time, usually five minutes or less.
- ❑ Grab sampling represents an environmental concentration at a particular point in time.
- ❑ Grab sampling is ideal following several phases of a cyclic process and for determining airborne concentrations of brief duration. It is seldom used to estimate eight-hour average concentrations.

- ❑ Grab sampling may require an evacuated flask or plastic bag for gas and vapor concentration analysis.
  - ❑ Grab samples introduce air into a container, which is sealed and then sent to a laboratory for analysis.
  - ❑ The laboratory may use gas chromatography, infrared spectrophotometry, or other methods to determine concentrations of gaseous contaminants.

- ❑ Grab sampling is rarely used alone to estimate an employee's eight-hour time-weighted-average exposure.

- ❑ Grab sampling can be used as a screening method to determine whether more extensive sampling is needed.

Air Sampling Instruments

AIR SAMPLING EQUIPMENT

- ❑ There are two categories for air sampling equipment:
  - ❑ Direct reading devices that provide immediate results.
  - ❑ Sample collection devices that collect a sample of air that is subsequently analyzed at a laboratory.

Sampling Train

- ❑ Air sample collection devices consist of five components. These components are collectively referred to as a sampling train and include:
  - ❑ An air inlet orifice.
  - ❑ A collection device.
  - ❑ An airflow meter.
  - ❑ A flow rate control valve.
  - ❑ A suction pump.

Operation of the Sampling Train

- ❑ Outlined is a description of the sampling train and its operation:
  - ❑ Air enters the sampling train through the orifice.
  - ❑ Airborne contaminants are collected on a collection medium, such as on a filter or in a liquid.
  - ❑ Airflow rate is set using the rate control valve.
  - ❑ Air flows through the device in liters or cubic centimeters per minute.
  - ❑ The suction pump moves air through the entire sampling train.

INSTRUMENT SELECTION

- ❑ Sampling for gases and vapors affords many choices of air-sampling instruments.

- ❑ Instrument selection depends on various factors, including type of contaminant under evaluation, whether direct reading equipment can provide instant results, and whether laboratory analysis will be required.

- ❑ Portability and ease of use are important for field use.

- Although some devices may be portable and easy to use, they may not be available. Therefore, availability of the equipment becomes important.
- The efficiency, reliability, and type of analysis of the device are important factors to consider.

Grab Sample Instruments

- Grab sampling is performed with evacuated containers of glass usually 200-1000 ml in volume, or plastic bags of Teflon, polyester, or polyvinylidene chloride.
- The correct container depends upon the reactivity of the gas and the container.
- Direct reading instruments are often used for grab sampling.
- Grab samples may be conducted with portable or fixed equipment.
- Grab sampling advantages include simplicity, inexpensiveness, and ability to collect 100% of the chemical.
- The major disadvantage of grab sample instruments is that reactive gases (such as hydrogen sulfide, sulfur dioxide, and nitrogen dioxide) cannot be collected because they may react with the container, atmospheric dusts, moisture, and other gases to produce poor results.

Sampling Pumps

- Integrated air sampling may require the use of sampling pumps. The sampling pump, or suction pump, is responsible for the movement of air through the sampling train.
- Pumps are generally lightweight, durable, quiet and operate on rechargeable nickel/cadmium batteries.
- Attach the pump to the worker's belt or pants.
- When considering a sampling pump, consider the airflow rate required, the pump's ease of use, and its suitability in a hazardous or flammable environment.
- Air sampling pumps are usually available in the various flow rate ranges including low flow pumps and high flow.

High and Low Flow Air Sampling Pumps

- Low flow pumps operate between 0.5–500 ml/min. The common flow rate for organic vapors is 200 ml/min. Low flow pumps are often used for solid sorbent tube sampling of inorganic vapors.

- High-flow pumps operate between 0.5–5 l/min, the common flow rate range for particulates, gases, and vapors. They are often used with collection filters, cyclones, and impinger sampling. The flow rate commonly used for dusts is 2.0 liters per minute (lpm, or l/min). Respirable crystalline silica is collected at 1.7 l/min.

Collection Processes

- An airborne contaminant is collected on media or in liquid media through two extraction processes, absorption and adsorption.

Absorption

- Absorption is the process of collecting a gas or vapor in a liquid collection medium. As the air stream passes through an absorption liquid, a gas or vapor is removed.
- The liquid media may be highly soluble and non-reactive with the gas or vapor. The liquid media may contain a reactive agent. The reactive agent reacts very quickly with the captured gas and forms a more stable compound.
- Liquid collection devices include gas wash bottles, spiral absorbers, and fritted bubblers. The gas wash bottle forces air through a nozzle into the absorbing liquid.

Impingers

- The midget impinger is the most commonly used type of gas wash bottle for collecting gases and vapors in liquid media. Place the impinger inside a holster and attach it to the worker's shirt collar.
- Spiral absorbers force air through the device in a spiral path through the liquid to increase the contact time between the air and the liquid and collect more gas.
- Fritted bubblers form tiny bubbles as air is forced through the fritted surface. This increases the surface area of air in contact with the absorbing liquid.
- Spiral and fritted bubblers are excellent for moderately soluble chemicals or those that react slowly with the absorbing liquid.
- Selecting an appropriate absorbing solution depends on the chemical nature of gas being collected. For example, Girard T reagent is the liquid medium for collecting formaldehyde.
- Flow-rates range from 0.25 to 3 liters per minute.

Adsorption

- Adsorption is the process of collecting insoluble or non-reactive gases using tubes filled with granular sorbent, such as activated charcoal, silica gel, porous polymers, or other materials.

- Attach small tubes containing sorbent materials to a sampling pump and place on the worker's shirt collar.

- As the pump pulls air through the device, the gas or vapor is adsorbed to the surface of the sorbent.

- Sorbents retain or adsorb gaseous substances chemically unchanged on their surfaces, which can later be desorbed (extracted) and analyzed at a lab.

- Activated charcoal and silica gel are the two most commonly used sorbents.

- Many solid sorbent materials with chemical coatings have been developed to sample reactive gases and vapors that cannot be collected on charcoal or silica gel. Examples include XAD-2, Tenac-GC, Ambersorb, and Chromosorb tubes.

Activated Charcoal

- Activated charcoal is the most widely used solid sorbent for organic vapor sampling. Characteristics of activated charcoal used in air sampling include:
  - Large surface area and reactive surface.
  - High adsorptive capacity for a large range of organic vapors.
  - Electrically non-polar; meaning it will preferentially adsorb organic vapors rather than polar molecules, such as water vapor.

- A standard charcoal tube is 7 cm long and 4 mm wide and divided into two sections. The first section contains 100 mg of charcoal, separated from the backup section. The backup or second section contains 50 mg of charcoal separated from the first section with fiber glass, glass wool, or a urethane foam plug.

- Some contaminants invariably pass through the first section. The backup section increases collection efficiency by adsorbing some of the material that was missed by the first section.

- Breakthrough occurs if gas or vapor in the backup section exceeds 10% of the mass in the front section. The calculated concentration is of questionable validity because a significant quantity of the contaminant may not have been collected.

- Activated charcoal is excellent for vapors boiling above 0° C. It is moderately useful for vapors boiling in the range -100°C to 0°C. It is poor for vapors boiling below -150°C.

- Activated charcoal tubes can be used to sample several types of vapors at once.

Passive Monitors

- Passive monitors allow personal sampling without the use of sampling pumps.
- The monitors operate through passive diffusion. The gas or vapor diffuse across a barrier into the monitor and are collected by a sorbent material.
- Since diffusion occurs at a specific rate, passive monitors rely on passive diffusion. The sample depends upon the diffusion coefficient of the organic vapor being sampled, the total cross-sectional area of the badge, and the length of the cavity (Fick's Law).
- Monitoring begins once the monitor cover is removed, time is recorded and the monitor is clipped into the breathing zone of the worker.
- Once monitoring is completed, the monitor is resealed and the time recorded before being sent to the laboratory.
- Determine mass by the amount of contaminant trapped on the badge. Calculate volume using the manufacturer's formula on the duration time of the sample.
- Passive vapor monitors are accurate within + or - 2.5% and are frequently used for a wide range of organic vapors, such as carbon monoxide, sulfur dioxide, ammonia, and formaldehyde.
- Passive monitors are easy to use and have no moving parts. They are small, lightweight, and inexpensive. They can monitor while being relatively unattended.
- Studies have demonstrated consistent results against charcoal tubes. Most meet or exceed the National Institute of Occupational Safety and Health (NIOSH) accuracy requirements.

Silica Gel

- Silica gel is another commonly used granular sorbent.
- Use silica gel for sampling inorganic substances such as acid gases or other electrically polar substances. Due to its polar nature, water vapor can seriously interfere with the adsorption of the contaminant.

- Use silica gel impregnated solid sorbent tubes for sampling reactive gases and vapors. The solid sorbents are impregnated with an absorbic reagent that changes the collected gas or vapor into a more stable ionic form. Once the sample is desorbed from the sorbent, analyze it with ion chromatography or other methods of analysis.

FLOW RATE METERS

- Sustaining a constant flow rate during sampling is extremely important. Sample pumps slow down if hoses get crimped or filters become dust laden.
- Failure to maintain adequate flow rate will adversely impact sampling results and interpretation of those results.
- Pressure compensating devices resolve the flow-rate variations commonly found in sampling situations. Pumps equipped with such devices detect pressure changes and maintain preset flow rates.

Critical Flow Orifice

- Some pumps use critical or limiting orifices to regulate airflow rate. This orifice is a precision drilled hole in a metal plate through which the air stream being sampled is directed.
- A critical flow orifice attached to a sampling pump allows the flow rate to remain constant despite changes in inlet conditions (e.g., clogged filter).
- When sampling pumps do not have a mechanism to maintain a constant airflow rate, use a calibration device, such as a precision rotameter, to check airflow during sampling and verify accuracy.

CALIBRATION

- Calibration of air-sampling equipment is a critical component of accurate data gathering.
- Temperature and pressure directly affect air volume. If conditions during air sampling are different from those during calibration, use a correction factor when calculating the field volume.
- Calibrate all sampling pumps before and after each day of sampling.
- If not performed on a primary standard, the secondary standard must have traceability to calibration against a primary standard.
- Prior to initial calibration before sampling, allow the pump to run about five minutes. Use the same sampling train during calibration when a sampling train requires an unusual combination of sampling media (e.g., glass fiber filter preceding impinger).

- Primary calibration standards include the frictionless piston meter or soap bubble meter, spirometer, Mariotti bottle, and electronic calibrators.

- Secondary calibration standards include wet test meter, dry gas meter, and rotameters.

Spirometers

- The spirometer is a primary volume standard that measures displaced air instead of displaced liquid.

- The spirometer operates as follows:
  - The spirometer contains a cylindrical bell with its open end under a liquid seal.
  - The weight of the bell is counter/balanced so that resistance to movement as air moves in or out of the bell is negligible.
  - Change in height times cross section measures the volume change.

Mariotti Bottles

- The Mariotti bottle is a primary volume standard that measures displaced water.

- The Mariotti bottle is a large glass jar with a valve at the bottom that, when opened, allows water to drain out of the bottle by gravity while air is drawn into the bottle to replace it. Volume of air drawn in is equal to the change in the water level multiplied by the cross section at the water surface.

Soap Bubble Meters

- Soap bubble meter is a primary calibration standard.

- The soap bubble meter, an air displacement measurement tool, is sometimes referred to as a "frictionless piston meter".

- The soap bubble meter may be used for high and low flow air-sampling pumps.

- Soap bubble meters are accurate within + or - one percent.

- The meter consists of an upturned laboratory burette (1000 ml for high flow and 100 ml for low flow pumps), ring stand and clamp, tubing, selected sampling media, and the pump.

- The soap bubble meter requires that the selected air sampling media be included in the sampling train during the calibration of the pump.

- The traditional upturned laboratory burette has been replaced by electronic bubble flow meters for pump calibration. Manufacturers of sampling pumps generally recommend their own brand of calibrator for their pump.

Electronic Calibrators

- Electronic calibrators have largely replaced many of the traditional primary calibration methods.
- Electronic calibrators provide instantaneous airflow readings and cumulative averaging of multiple samples.
- These calibrators measure the flow rate of gases and present the results as volume per unit of time.
- Periodically check batteries. Alkaline batteries should be replaced monthly and rechargeable batteries must be checked periodically under load.
- These units are highly accurate and portable.
- Electronic calibrators can clog if soap is too thick, and Nylon adapters can restrict airflow due to plugging.

Secondary Standards

- Secondary standards rely on periodic recalibration against primary standards. They are accurate within + or - one percent.
- They include wet-test meters, dry-gas meters, and precision rotameters.

Wet Test Meters

- A wet test meter is a secondary standard.
- The wet test meter consists of a partitioned drum, half submerged in a liquid (water), with openings in the center and periphery of each radial chamber.
- Air enters at the center and flows into one compartment, causing it to rise. This causes rotation.
- The number of revolutions is indicated on a dial on the face of the meter.
- Liquid is displaced by air. The volume measured depends on the height of the fluid in the meter.
- These meters are accurate within one percent.
- Errors associated with these meters are due to moving parts that are subject to:
  - Corrosion.
  - Friction in the bearings.

- Other errors may be caused by the fact that:
  - Inertia must be overcome at low flows (< 1 RPM).
  - High flow liquids (> 3 RPM) might surge and break the water seal at inlet or outlet.

Dry Gas Meter

- Dry gas meters are similar to domestic gas meters and are advantageous under high pressures and volumes.
- They consist of two bellows connected by mechanical valves and a counting device. As air or gas fills one bag, it mechanically empties the other.
- Disadvantages include mechanical drag, pressure drop, and leakage.

Precision Rotameters

- Precision rotameters are secondary standards. These units are attached to high-flow sampling pumps.
- Precision rotameters consist of a float and ball free to move up or down within a vertical tapered tube that is larger at the top than the bottom.
  - Air is pumped into the bottom of the tube.
  - The ball will rise until an equilibrium is reached between the weight of the ball and the force exerted on the ball by pressure of air flowing through the annular area between the ball and the wall of the tube.
  - The higher the ball, the faster the flow rate.
  - The height of the ball is measured against a numerical scale attached to the tube and is an indirect measure of flow-rate.
- They are very accurate within + or - one to five percent, unlike rotameters built into a sampling pump.
- They can be easily placed in a sampling train to calibrate the pump before or after the sampling period, or during use of the sampling pump.
- Calibrate precision rotameters against a primary standard (typically a bubble meter) at regularly scheduled intervals.

Sampling Methods

SELECTION OF SAMPLING METHODS

- Selection of an air sampling method depends upon the sampling objective and whether it involves compliance or documenting exposures.
- Physical and chemical characteristics of the contaminant are important when determining the sampling method.
- If other chemicals are present, they may interfere with collection and/or analysis.

- Required accuracy and sensitivity should be reviewed before determining a sampling method.
- If regulatory requirements are an issue, they may prescribe a particular sampling or analytical method.
- The complexity of the method and the unit's portability are critical when working in the field. The more portable, the better.
- The cost and reliability of the sampling method or analytical method can affect sampling method selection.
- The type of sample needed (area vs. personal) may determine the sampling method.
- The duration of sampling may determine the method selected.
- Professional judgment and personal preference may play a role in sampling method selection. For example, few industrial hygienists choose impinger sampling if an alternative method is available.
- For assistance, consult with the laboratory that will be used to calculate the sample. This assures the method selected meets the analytical ability of the laboratory. Only AIHA certified laboratories should be used.

Standardized Sampling Methods

- All of the information necessary to sample the air for a specific contaminant is provided through standardized sampling methods.
- Analytical procedures specify the collection media, sample volume, flow rate, and chemical analysis.
- These procedures are based on:
  - NIOSH's *Manual of Analytical Methods.*
  - American Public Health Association's (APHA) *Methods of Air Sampling and Analysis.*
  - OSHA's *Chemical Information Manual.*
  - The laboratory's experience.
  - EPA guidelines and regulations.
- The laboratory used must be accredited by the American Industrial Hygiene Association (AIHA) and participate in the NIOSH Proficiency Analytical Testing Program. These two organizations monitor laboratory performance to ensure accuracy.

5 Direct-Reading Instruments for Air Sampling of Gases, Vapors, and Particulates

Direct-reading instruments can measure the concentration of many gases, vapors and particulates in air. These instruments perform both sampling and analysis. An industrial hygienist must be familiar with the uses, advantages, and limitations of the various direct-reading instruments used to evaluate workplace hazards.

Colorimetric Devices

DIRECT-READING COLORIMETRIC DEVICES

- ❑ Direct-reading colorimetric devices use the reaction of an airborne contaminant with a color-producing agent to reveal a stain length or color intensity.
- ❑ Stain lengths or color intensities can be directly read to provide an instantaneous value of the concentration.
- ❑ These devices are widely used, simple to operate, and inexpensive.
- ❑ Colorimetric devices are versatile in the detection of numerous contaminants and accurate within + or - 25%.
- ❑ Observer variability, or how the stain length or color intensity is read, causes some variability in results.

Flow Rates

- ❑ Colorimetric tubes in an active system require the use of a vacuum pump. The pump is usually a manual vacuum that is generated through bellows or by pulling air through the pump.
- ❑ Check flow rates regularly on such vacuum pumps. Airborne dusts and lint can clog orifices. Maintain vacuum pumps consistent with the manufacturer's recommendations.

Passive System

- ❑ A passive colorimetric system does not require a vacuum pump.
- ❑ A passive system works by diffusion, is simple and economical.
- ❑ Insert a passive colorimetric detector tube or device tube into a holder and attach to worker's clothing. Passive colorimetric badges work in the same manner.

Length of Stain Detector Tubes

- ❑ A length of stain colorimetric device measures the length of stain produced, or the ratio of stain length to gel length compared against a chart to determine the percentage of concentration.

Color Change Detector Tubes

- ❑ A color change detector tube colorimetric device is similar to a length of stain colorimetric device.
- ❑ The difference is that a uniform color is produced in the reactive section when the contaminant gas or vapor reacts with the indicating reagent in a color change device.
- ❑ Compare color to the standard color chart that is provided with each instrument of this type.
- ❑ Disadvantages:
  - ❑ Colors fade with time.
  - ❑ Requires visual judgment.
  - ❑ Depends on color perceptions and lighting conditions.

Shelf-Life

- ❑ Specific tubes may not be used often and may not be exhausted before the manufacturer's indicated expiration date.
- ❑ Refrigeration extends shelf life. Warm before using.

Specificity

- ❑ Each system measures a specific gas or vapor. They are subject to interference that can invalidate the sample.
- ❑ Many common gases react with the same chemicals or have similar physical properties.
- ❑ Some indicator tubes need a preconditioning system to remove potentially contaminating substances. They react with the gas to convert the compound to a more suitable reacting compound. They react with the new gas or vapor released and are measured by the second section.
- ❑ Refer to the manufacturer's list of interfering substances and conditions for specific detector tubes.
- ❑ Use detector tubes at room temperature.
- ❑ Interchanging tubes obtained from various manufacturers will lead to erroneous results. Sampling rates of various pumps may not be the same, as well as variations in the reaction rate of chemicals in the tubes.

Colorimetric Tape Samplers

- Colorimetric tape samplers are chemically impregnated paper tapes. They have been used since 1950s and were first developed for hydrogen sulfide.
- Colorimetric tape samplers consist of a lead acetate reagent system impregnated into filter paper that generates a dark stain (lead sulfide) when exposed to hydrogen sulfide. The concentration is measured by the transmission of light through paper.

Principles of Operation

- Chemically treated paper tape is drawn over the sampling orifice. Then:
  - Contaminant reacts with the chemical to produce a stain.
  - Intensity of the stain can be measured by a reflectometer.
  - Tape moves across a reel.
  - Constant flow of air is aspirated through porous tape by self-contained pump.

Electrical Direct-Reading Instruments

COMBUSTIBLE GAS METERS

- Various manufacturers make combustible gas meters with different designs. Portable and stationary designs exist.
- Some combustible gas meters are combined with oxygen meters. Most have a length of tubing with a probe on the end for confined spaces, etc.
- Before use, a brief warm-up period is required to heat filaments.
- Use in an atmosphere known to be free of combustibles when making zero adjustments on the meter. Interference is possible.

Combustible Gas Meter Operation

- Each instrument contains a battery-operated electrical circuit called a Wheatstone bridge circuit. These instruments operate in the following way:
  - A measurable amount of heat is released when a combustible gas or vapor is burned.
  - On part of the bridge, air is passed over filaments that have been brought to a high temperature.
  - If air contains a combustible gas or vapor, heated filaments cause combustion and additional heat is released. This increases the electrical resistance of the filaments.
  - Another part of the bridge contains sealed filaments. These are also heated, although not in the air stream.
  - Sealed filaments cancel out all changes in electrical current and resistance due to temperature variations in the wire.
  - Net effect: the change in resistance to the electrical current flow in the filaments in the air stream is due to the presence of combustible gases.

- Changes in current are registered as "percent LEL" on the instrument panel:

  Δtemperature → Δresistance → current flow → meter deflection

Explosive Limits

- Upper explosive limit (UEL) is the maximum volume percentage of the material that can be ignited. If the UEL is exceeded, the mixture cannot be ignited and sustain combustion, since there is insufficient oxygen present to support combustion.

  Lower explosive limit (LEL) is the lowest percentage at which an explosion or fire can occur. Mixture is too lean to ignite.

CARBON MONOXIDE METERS

- The most common carbon monoxide instrument uses a potentiometric or coulometric cell relying on a change in voltage difference for measurement.
- Use in parking garages, bus terminals, sewers, vaults, and mines.

Carbon Monoxide Meter Operation

- Carbon monoxide meters use electrochemical sensors with potentiometric or coulometric cells.
- The cell electro-oxidizes carbon monoxide (CO) to produce carbon dioxide in proportion to the partial pressure in the sample area. The resulting electrochemical signal is amplified and the temperature is compensated to drive the meter.
- Carbon monoxide meters may have a battery-operated pump or operate through diffusion.
- Portable units contain visual and audible alarms. Their range is from 1–500 ppm.

OXYGEN METERS

- Use oxygen meters in enclosed spaces, manholes, tunnels, mines, or other confined spaces.
- Twenty-one percent oxygen in air is normal. Sixteen percent is the minimum concentration to support life. Consider an atmosphere to be oxygen deficient at nineteen and one-half percent. When oxygen increases in the atmosphere to twenty-five percent, the atmosphere can be explosive.

Oxygen Meter Operation

- Galvanic sensing cells measure oxygen on either the 0-25% or 0-100% by volume range. A minute electrical current is generated in the cell proportionate to the oxygen content in the atmosphere.

- Oxygen is paramagnetic (attracted to a magnetic field). Magnetic lines of flux pass through oxygen more easily than through other gases. When oxygen is heated, it loses its paramagnetic properties, or is repelled out of a magnetic field. During operation, a sample of air passes through the oxygen analysis cell over an electrically heated resistor. As it passes over the resistor, it is heated and loses magnetic properties. The cooling effect of the resistor is proportional to the oxygen content of the gas.

- Most oxygen meters are accurate and simple.

CHLORINE AND HYDROGEN SULFIDE METERS

- Chlorine and hydrogen sulfide meters are similar to carbon monoxide meters and use electrochemical sensors. They may operate by diffusion.

- Use hand-held or belt-worn units when entering confined spaces. Most contain visual and audible alarms.

- These meters are subject to interference.

INFRARED ANALYZERS

- When they absorb infrared radiation over a spectrum of wavelengths, many gases have a characteristic infrared spectra that that can be converted into characteristic graphs. Use an infrared analyzer to detect the presence of air contaminants and determine concentrations.

- The MIRAN by Foxboro is one example of an infrared analyzer. These units are portable and highly sensitive.

Operation of Infrared Analyzers

- An infrared analyzer operates as follows:
  - An infrared generator in the analyzer emits the full frequency range.
  - Window material in the cell limits the frequencies that can be used (wavelengths from 0.8 to 200 micrometers).
  - The source simultaneously emits infrared energy through two cell paths.
  - The opposite end is a detector that monitors the energy being transmitted through the two cells.
  - One cell is the sample cell. Gas is injected into the sample cell, then analyzed.
  - The other cell is sealed and is the comparison cell. If the gas injected into the sample cell contains a gas that absorbs energy at the selected frequency, the detector will detect less energy coming through the sample cell than the comparison cell.

- ❑ The detector emits an electrical signal to alert the user to this imbalance.

PHOTO-IONIZATION DEVICES

- ❑ A photoionization device (PID) is a direct reading instrument that operates by detecting and distinguishing between ions of vapors and gases following ionization by the instrument's ultraviolet light source.

- ❑ Although these units originated in chemical laboratories, they are now available in portable models.

GAS CHROMATOGRAPHY

- ❑ Gas chromatography is a gaseous detection technique that involves the separation of mixtures by passing them through a column that enables the components to be retained for varying periods of time before they are detected and recorded. Each chemical has a specific retention time.

- ❑ Use gas chromatography (GC) to analyze complex chemical mixtures.

- ❑ Gas chromatographs are widely used, reliable, lab based or portable, easy to use, and easy to calibrate.

- ❑ A gas chromatograph consists of:
  - ❑ A carrier gas with a pressure regulator or flow meter.
  - ❑ Injection system for gas samples.
  - ❑ Separation columns of stainless steel, glass, or copper.
  - ❑ Heater/oven assembly.
  - ❑ Detector.
  - ❑ Recorder for chromatograms.

Gas Chromatography Operation

- ❑ Gas chromatography depends upon differential migration of the components through the porous medium of the column, usually diatomaceous earth.
  - ❑ Movement through the column depends on the chemical affinity for the column material.
  - ❑ Differential migration is carried out so that each component separates as a discrete substance.
  - ❑ Each compound in the mixture will be uniquely retained by the stationary phase as it travels through the packed column.
  - ❑ Each gas emerges from tube at a unique time. The height and area of peak are proportional to the concentration of the eluted sample component.
  - ❑ Emerging gases may then be detected using flame ionization detection (FID), electron capture (EC), etc.
  - ❑ A particular reference substance and the sorption time sequence are used for identification and detection of an isolated gas in a mixture.

- ❑ A readout system includes a thermal desorption unit for removal of the collected sample from charcoal tubes.

❑ Types of gas chromatography include:
- ❑ High performance liquid chromatography (HPLC).
- ❑ Liquid chromatography (LC).
- ❑ Ion Chromatography (IC).

FLAME IONIZATION DETECTOR (FID)

❑ A flame ionization device (FID) is a direct reading instrument that ionizes gases and vapors with an oxyhydrogen flame and measures the differing electrical currents generated.

❑ Flame ionization offers sensitive detection of various hydrocarbons. FID responds to all organic compounds except formic acids.

❑ Flame ionization increases production of ions by introducing a volatile carbon compound into a hydrogen flame. A loop of platinum held above the flame serves as the collector electrode. A current carried across the electrode gap is directly proportional to the number of ions produced. Use in combination with gas chromatography.

PHOTOMETRY DEVICES

❑ Photometry measures the relative radiant power of a beam of radiant energy in the visible, ultraviolet, or infrared region of the electromagnetic spectrum.

❑ The most sophisticated form of photometry is spectrophotometry. Spectrophotometry uses prisms made of glass (visible region), sodium chloride (UV), and potassium bromide (infrared) to provide monochromatic radiation, a "purer" source of energy.

❑ These units are used primarily in laboratories, are highly specific, and very precise.

Aerosol Photometers

❑ An aerosol photometer is a device used to approximate particle size classification.

❑ An aerosol photometer operates as follows:
- ❑ Generation of electrical impulse by photocell.
- ❑ Detects light scattered by a particle.
- ❑ Pulse height analyzer estimates particle diameter.
- ❑ The number of electronic pulses corresponds to the number of particles of sampled gaseous medium.

Spectrophotometers

- Spectrophotometry covers a wide variety of sampling and analytical technologies.
- A spectrophotometer is a direct reading instrument used for comparing the relative intensities of corresponding electromagnetic wavelengths produced by absorption of ultraviolet, visible, or infrared radiation by a vapor or gas.
- Examples include:
  - Visible spectrophotometry.
  - UV spectrophotometry.
  - Infrared spectrophotometry.
  - Atomic absorption spectrophotometry.
  - Emission spectrophotometry.

Chemiluminescence Monitors

- Chemiluminescence occurs with certain chemical reactions. It is a distinct colored glow that accompanies the chemical reaction (e.g., Luciferin oxidation in fireflies).
- Chemiluminescence units are incorporated into continuous ambient air monitors for ozone. Examples: Ozone is determined to be in an air sample stream with ethylene gas or nitrogen oxide based on the chemiluminscent reaction of nitrogen oxide and ozone to produce nitrogen dioxide (NO_2).

COULOMETRY

- Coulometry measures the number of electrons that are transferred across an electrode solution interface when a reaction in the solution is created and carried to completion. The reaction is caused by a contaminant in a sample gas that is drawn through or onto the surface of the solution. The number of electrons transferred, measured in coulombs, is an indication of the contaminant concentrations.
- Use for reactive inorganic gases present in ambient atmospheres, ozone, nitrogen dioxide (NO_2), and sulfur dioxide (SO_2).
- These instruments are very precise and measure the number of electrons in terms of coulombs.

ELECTRICAL CONDUCTIVITY DECTECTORS

- Electrical conductivity occurs when a gas-air mixture is drawn through an aqueous solution. Those gases that form electrolytes produce a change in electro-conductivity as a summation of the effects of all ions produced.
- Use the increase in conductivity as a measure of concentration.

- The greatest application is in sulfur dioxide (SO_2) continuous ambient air monitors and in portable SO_2 monitors where SO_2 in the presence of peroxide forms sulfuric acid (H_2SO_4).

THERMAL CONDUCTIVITY MONITORS

- Thermal conductivity measures specific heat of conductance of a gas or vapor as a measure of its concentration in a carrier gas (Ar, He, H, N).

- This is useful in estimating the concentration of the separately eluted components from a GC column. Detectors register a loss of heat from a hot filament to a single component of a flowing gas stream. Loss is registered as a decrease in electrical resistance measured by a Wheatstone bridge.

CALIBRATION

- There are two common methods for calibrating direct-reading instruments:
  - The static method—A known volume of gas is introduced into the instrument and sampling is performed for a limited period of time.
  - The dynamic method—A known concentration of the contaminant is prepared in a test chamber and the instrument is used to monitor that concentration.

- The static method is easier and more efficient, since the dynamic method requires facilities not always available in a workplace.

6 Toxicology

Understanding the many factors affecting toxicity is critical knowledge for the industrial hygiene professional. A substance's ability to cause bodily harm depends on duration of exposure, method of entry into the body, concentration of contaminant, and individual risk factors, among other variables. Familiarity with the principles of industrial toxicology enables industrial hygienists to minimize workplace exposures and protect workers.

Basic Principles

TOXICOLOGY AND TOXICITY

- Toxicology is the study of a substance's ability to cause damage to the body.
- Many factors affect how toxic a material may become, including:
  - How the substance enters the body (Routes of Exposure).
  - The amount of toxin taken in the body (Dose).
  - The frequency and duration of exposure.
  - The site of action within the body (Target Organ).
  - The age, weight, and health status of the exposed person.
- A poison is a substance that through its chemical action kills, injures, or impairs an organism: a substance that inhibits the activity, course of a reaction, or process of another substance.
- Risk is the probability that a substance will produce harm under specified conditions of use.
- Safety is the probability that harm will not occur under specified conditions of use.
- Signs are objective evidence of disease to the examining health professional.
- Symptoms are subjective evidence of disease perceived by the patient.

LD50/LC50

- The terms LD50 and LC50 represent lethal dose (LD) and lethal concentration (LC) for 50% of test animals included in a scientific study.
- LD50s are expressed in a unit of weight of a substance per kilogram of animal body weight (e.g., mg/kg).
- When considering inhalation exposures, the LC50 is used with results expressed in terms of an air concentration (e.g., ppm).

- LD50 and LC50 data should always be accompanied by a notation of the species of animal, the route of exposure (for LD50), such as IV, oral, etc., and the length of time over which the animals were observed (generally 14 days).

Significant Toxins and Effects

- Vinyl chloride—angiosarcoma. Vinyl chloride is actually metabolized to an epoxy derivative. It is this epoxy derivative that is thought to be the proximate carcinogen.
- Asbestos—mesothelioma, asbestosis.
- PCBs—chloracne, liver impacts, bioaccumulates.
- Thalidomide—teratogen (phocomelia—absence of limbs in offspring).
- Methyl alcohol—blindness when ingested.
- Mercuric nitrate—central nervous system damage ("mad as a hatter").
- Silica—silicosis (fibrosis).
- Isocyanates (TDI, MIC)—sensitization (allergic type reaction).
- Cotton dusts—byssinosis (tightness in chest, chronic bronchitis).
- Organophosphate insecticides—acetylcholinesterase inhibitors.
- Organochlorine insecticides (DDT, chlordane)—neurotoxins.
- Carbon monoxide—carboxyhemoglobin formation (CO binds strongly with hemoglobin).
- Cyanide—used in capital punishment of criminals (gas chamber). Blocks the use of oxygen at the cellular level (inhibits cytochrome oxidase).
- Methylene chloride—metabolized to carbon monoxide.
- Tri-otho-cresol phosphate (TOCP)—Ginger Jake paralysis. Produced dreadful epidemics when sold as ginger in 1930s in southern US. Causes peripheral neuropathy.
- Paradichlorobenzene—modern day mothballs, deodorizers, seed protectants.

Classification of Toxic Agents

- Toxic agents are classified in a variety of ways depending on interests and needs.
- Toxic agents are discussed in terms of their:
  - Target organ (liver, kidney, hematopoietic system, etc.).
  - Use (pesticide, food additive, etc.).
  - Source (animal and plant toxins).
  - Effects (carcinogens, mutagens, etc.).

Forms of Toxic Compounds

- Exposures to toxic compounds can be in the form of liquids, gases, mists, fumes, dusts, and vapors.
- Gas—States of matter in which molecules are unrestricted by cohesive forces. They expand to occupy the space in which they are confined.
- Vapors—The volatile form of a substance that is normally in the liquid or solid state at room temperature. Vapors are generated through evaporation.
- Fumes—Formed when the material from a volatilized solid condenses in cool air. The solid particles that are formed are generally extremely fine (usually less than one micrometer). In most cases, the volatilized solid reacts with air to form an oxide. Example: welding fumes.
- Mists—Suspended liquid droplets. May be the result of condensation of a liquid vapor or the breaking up of liquids by splashing or atomizing. Example: spray painting.

ROUTES OF EXPOSURE

- Toxic materials can enter the body through various routes.
- The three primary routes of exposure, in order of importance, are inhalation, skin absorption, and ingestion.
- Other routes, generally less important from an occupational standpoint, are the injection (or parental) routes such as intravenous (IV), subcutaneous (SC), intramuscular (IM), and intraperitoneal (IP). These routes of exposure are of interest from an occupational standpoint based on skin puncturing, particularly in the medical field (sharps). However, they are all used extensively in toxicity testing, particularly the IP route.
- The route of exposure can obviously influence the concentration of a toxic material that reaches the site of action. In that respect, the route of exposure results in differences between dose and exposure.

- The only time exposure equals dose is in the case of injection (particularly IV). Otherwise, factors such as incomplete absorption will result in exposure exceeding dose.

- Some substances are very toxic but because of their properties (such as low vapor pressure, poor skin absorption, etc.), they do not present a particular hazard in a workplace situation.

Inhalation

- Toxic responses to chemicals can result from their absorption through the lungs (example—a recently painted room can cause a headache). Other common examples include carbon monoxide poisoning, asbestos, hydrogen cyanide (gas chamber), and chemical warfare agents (mustard gas, etc.).

- The effects of different substances following inhalation depend on particle size and gas solubility.

- A primary factor that limits the amount of material absorbed through the lungs is the rate of respiration. The amount of material absorbed can increase significantly if the rate and depth of respiration increases.

- The average person breathes in about one cubic meter of air per hour at rest. With a moderate activity level during work, the amount of air breathed in will increase through shifting from nasal breathing to oral-nasal breathing, thereby increasing the potential amount of toxin entering the body.

Particle Size

- Only small particles (generally less than 5-10 micrometers) can traverse through the respiratory system to be deposited in the gas exchange areas of the lungs (called alveoli).

- The lung protects itself through several mechanisms including nasal hairs, mucociliary clearance, and impaction of particles through the treacherous bifurcations in the lung. Macrophages also collect particles.

Solubility

- The solubility of a gas will determine whether it is absorbed in the upper respiratory system (mucous). The solubility will also determine the extent to which a gas will be absorbed by the blood if the gas makes it to the alveoli.

- Highly soluble compounds are stopped in the upper respiratory system. Examples include irritants like ammonia, hydrogen chloride, and sulfur dioxide.

- Intermediately soluble compounds are considered whole respiratory irritants and include chlorine, bromine, and ozone.

- Less soluble compounds are lower respiratory irritants, including nitrogen dioxide and phosgene.

Skin Absorption

- Healthy, undamaged skin acts as a natural barrier to many substances.

- However, some substances are readily absorbed through the skin in sufficient amounts to produce either local or systemic toxic effects. This absorption can occur through:
  - Abrasions.
  - Sweat glands.
  - Sebaceous glands (oil glands).
  - Hair follicles.
  - Directly through epidermal cells.

- Some examples of chemicals that readily penetrate the skin are solvents (e.g., carbon tetrachloride), insecticides, mercury, and PCBs.

- Skin contact can result in varying toxic reactions:
  - Local irritant (acids and bases).
  - Sensitization (chloracne, photosensitization).
  - Systemic poisoning.

Ingestion

- Ingestion, the most important route of exposure in cases of suicide and childhood poisoning, is generally less important in occupational settings.

- Eating meals or smoking in an area where food and cigarettes can become contaminated can increase the level of toxins absorbed through ingestion.

- Poor personal hygiene (e.g., not washing hands after working in a contaminated environment) can also increase the amount of toxin ingested.

FACTORS AFFECTING TOXICITY

- Concentrations of contaminant, length of exposure, and duration of exposure are three significant factors affecting toxicity.

Duration of Exposure

- The longer the exposure, the greater the potential that the substance will exhibit a toxic effect in the body. Longer exposures also provide the body with less time to recover between exposures. A "normal" workday with eight hours of exposure means that the body has 16 hours to recover from that exposure by metabolizing and excreting the substance. If the workday is longer than eight hours, there is proportionally less time for the body to recover.

Frequency of Exposure

- Frequency of exposure relates to the body's ability to recover. Off the job exposures also impact the body's ability to recover.

Absorption

- If a chemical is not absorbed well following exposure (inhalation, skin contact, and ingestion), the chemical may prove to be much less hazardous in the work environment than what animal toxicity data may suggest.
- This is particularly true if the toxicity data is based on intravenous (IV) application.

Rate of Metabolism

- The rates of metabolism and excretion affect toxicity.
- Certain chemicals are easily metabolized and excreted by the body.
- Other substances tend to accumulate in certain organs of the body and are not readily cleared. These chemicals tend to develop a "body burden" (e.g., PCBs and lead have low acute toxicity but higher chronic toxicity).

Individual Susceptibility

- Reactions or responses to toxins vary depending on individual susceptibility. Some workers may be hypersusceptible or more sensitive than others. This is particularly true with allergens.
- Individual susceptibility is one of the most difficult things to control and to anticipate. It may be a permanent condition or the result of a temporary condition such as a cold or asthma.

Latency Period

- Many toxic effects do not show up immediately after exposure. There may be a short or lengthy latency period between exposure and the onset of disease.
- Example: Cancer from asbestos exposure may take 20 to 30 years after exposure for symptoms to develop.

Synergistic Effects

- Synergistic effects may occur when a person is exposed to more than one toxic substance at a time, or when one exposure follows the other in a short time period.
- One of the substances increases the effect of the other. The resulting toxicity is much greater than the sum of the individual toxicities.
- An example of synergistic effects is two chemicals that are metabolized by the same enzyme, but one chemical is metabolized preferentially. Therefore, the less metabolized chemical may be metabolized much more slowly than normal, or not at all, which therefore increases toxicity.
- Examples of synergistic effects:
  - Smokers have a 10 times greater risk of developing cancer than non-smokers.
  - Non-smokers exposed to asbestos have a six times greater risk of developing lung cancer than non-smokers with no exposure to asbestos.
  - Smokers who are also exposed to asbestos have a fifty to ninety times' greater risk of developing cancer than the non-smoking, non-exposed population.

Age and Gender Factors

- The very old and the very young may be more susceptible to toxic effects than the healthy working population. This is partly due to an impaired immune system, and immature or damaged livers.
- Some chemicals can have specific effects on the reproductive system of men or women.
- Examples of age and gender specific responses include:
  - Dibromochloropropane (DBCP)—A pesticide that results in abnormal sperm production with resultant decrease in the ability of men to get their wives pregnant.
  - Ethylene dibromide (EDB)—A lead scavenger in leaded gasoline and a pesticide that has the effect of causing testicular damage.
- Other factors could relate to body fat (women tend to have higher percentage body fat than men).

Acute Effects

- Acute exposures and acute effects generally involve short-term, high concentration exposures, and immediate results of some kind (illness, irritation, and death).

- The effect of a chemical exposure is considered acute when it appears with little lag time, such as within seconds, minutes, or hours.
- Examples: Inhaling high levels of carbon monoxide will produce acute poisoning. Another example is the irritation caused by the smoke of a fire.
- Acute toxicology tests generally involve a single dose, 14-day follow-up. Scientists look for overt signs such as serious illness, weight loss, etc. Acute toxicology tests are the basis for the development of LD_{50} data.

Chronic Effects

- Chronic effects relate to the continued exposure to a substance.
- Chronic poisoning assumes that some level of the substance will be continuously present in the body.
- The symptoms of chronic poisoning from a chemical are usually different than the acute symptoms from the same chemical.
- Examples of chemicals that may produce chronic health effects include:
  - Benzene's acute effect is dizziness, but low level chronic exposure may result in effects in damage to the blood forming organs (e.g. leukemia) or liver.
  - Hydrogen sulfide has an acute action on the nervous system that results in respiratory paralysis. With chronic, low-level exposure, hydrogen sulfide is rapidly oxidized to relatively non-toxic substances.
- Chronic toxicology tests should be conducted over the lifetime of the subject/animal (e.g., two years for rats and mice). Scientists look for cancer, biochemical changes, reproductive effects, etc.

Dose-Response Relationship

- Describes the relationship between particular dose of a substance and its corresponding reaction (response).
- Toxicity Testing: Response is often death (lethality). Dose response curves are generated to determine the LD50 of a chemical.

Body Systems Commonly Affected by Toxins

RESPIRATORY SYSTEM

- Humans can survive for weeks without food, days without water, but only a few minutes without air. Oxygen, the life-giving component of air, constitutes less than 21% of air.

- ❑ Respiratory system:
  - ❑ Starts as a tube or airway (trachea).
  - ❑ Becomes progressively smaller in diameter.
  - ❑ Terminates in small pouches known as alveoli.
  - ❑ Alveoli resemble small bunches of grapes.
  - ❑ Lungs receive blood from the heart.
  - ❑ In the alveoli, blood picks up oxygen from the air.
  - ❑ Removes carbon dioxide (due to pressure differentials).

- ❑ The uptake of toxic gases can occur throughout the respiratory system, unlike the exchange of oxygen and carbon dioxide, which occurs in the alveoli.

Toxic Responses of the Lungs

- ❑ Toxic responses of the lungs can include irritation, sensitization, scar formation, cancer, pneumonia, and emphysema.

- ❑ Irritation. Constriction of air passages.

- ❑ Sensitization. Allergic reaction leading to airway constriction (e.g., isocyanates).

- ❑ Fibrosis. Production of fibrotic tissue results in decreased surface area for gas exchange and decreased elasticity, which makes expiring air difficult. Difficulty in expiring air can increase carbon dioxide levels in the alveoli and decrease diffusion of oxygen into the blood. Examples include silica, asbestos, and beryllium.

- ❑ Carcinogenesis. Lung cancer (smoking, asbestos, arsenic).

- ❑ Pneumonia. An outpouring of fluids into the lungs in response to irritation or infection. Results in decreased gas transport.

- ❑ Emphysema. Symptoms include decreased elasticity of the bronchioles and the breakdown of alveolar walls, resulting in several alveoli joining together to form one large alveoli—but with less surface area for gas exchange.

SKIN

- ❑ Skin is the largest organ of the body and comprises about 2800 square inches of surface area.

- The skin's functions include:
  - Protecting the body against bacteria.
  - Protecting the more sensitive tissues of the body.
  - Protecting against the sun's rays.
  - Protecting against the loss of moisture.
  - Serving as an organ of perception for the nervous system.
  - Serving as a temperature regulator.

Layers of the Skin

- The skin is made up of three layers of tissue:
  - Epidermis—consists of several layers, including the oily layer (or acid mantle) and the horny layer (dead cells).
  - Dermis ("true skin")—contains blood vessels, nerves, hair follicles, sweat glands, and oil glands.
  - Subcutaneous—fatty tissue that provides elasticity and cushion.

Types of Skin Reactions

- Industrial dermatitis continues to be a major form of job related disease.

- Industrial dermatitis is a general term used to designate inflammation of the skin resulting from exposure to irritants in the work environment. Principal symptoms are redness, swelling, sensation of heat, and itching.

- There are four general types of toxic skin reactions: primary irritation, sensitization dermatitis, photosensitization, and skin cancer.

- Primary irritation can be caused by mechanical agents (friction), physical agents (extreme heat or cold), and chemical agents (acids, alkalis, irritant gases and vapors). Primary irritants may result in inflammation without allergens.

- Sensitization dermatitis may result in an allergic reaction.
  - First exposure may not cause irritation.
  - Extended or repeated exposures in some people result in allergic skin reaction.
  - Once a person has become sensitized, even a slight exposure can cause severe dermatitis.
  - Examples include nickel and chromium.

- Photosensitizers are chemicals that increase the skin's sensitivity to sunlight. Examples include tetracycline and creosote.

- Carcinogenesis. Examples include UV light, and polycyclic aromatic hydrocarbons.

CENTRAL NERVOUS SYSTEM

- The central nervous system (CNS) is protected from many toxic agents by the blood-brain barrier (BBB), but not all substances are blocked by the BBB. Some anesthetics, analgesics, and tranquilizers penetrate easily.
- Other chemicals that can cross the BBB and cause damage in the CNS include:
  - Alcohols—can cause nerve damage.
  - Carbon disulfide—can cause psychosis and tremors.
  - Organomercury compounds—e.g., methyl mercury.
- Development of the BBB occurs over time and may result in different effects in children vs. adults. For example, inorganic lead salts may accumulate in the CNS of children, while in adults the effects of exposure are generally seen in the peripheral nervous system. This explains why lead is more dangerous to children.

EXCRETORY SYSTEM

- The primary functions of the kidneys are the excretion of wastes and the control of salts (electrolytes).
- General toxic effects of chemicals on the kidney include decreased elimination of wastes with a resultant increase of waste materials in the blood (e.g., Blood urea nitrogen, BUN).
- Specific kidney toxicants include:
  - Heavy metals—mercury (particularly the inorganic forms), chromium, uranium, and cadmium.
  - Halogenated hydrocarbons.
  - Chloroform ($CHCl_3$).
  - Carbon tetrachloride (CCl_4).

Toxicity Rating Chart

Adapted (and subsequently revised) from Cassarett and Doull's *Toxicology: The Basic Science of Poisons*, Second Edition.

Probable Oral Lethal Dose for Humans			
Rating	**Dose**	**Average Adult**	**Example**
Practically nontoxic	>15 g/kg	More than 1 quart	Ethyl alcohol
Slightly toxic	5–15 g/kg	Bet. 1 pint & 1 quart	Ethyl alcohol
Moderately toxic	0.5–5 g/kg	Ounce–pint	Morphine sulfate
Very toxic	50–500 mg/kg	Teaspoon–ounce	Phenobarbital sodium
Extremely toxic	5–50 mg/kg	7 drops–teaspoon	Strychnine sulfate
Supertoxic	<5 mg/kg	A taste (<7 drops)	Botulinus toxin

Carcinogens

TYPES OF CARCINOGENS

- Chemical carcinogens are defined operationally by their ability to induce tumors. Carcinogen literally means "giving rise to cancer."

- Two major classifications of carcinogens exist: genotoxic and epigenetic.

GENOTOXIC CARCINOGENS

- Genotoxic carcinogens are those that interact with and alter DNA. These can be identified through short-term *in-vitro* tests.

- Genotoxic carcinogens include:
  - Direct-acting or primary carcinogens.
  - Pro-carcinogens or secondary carcinogens.
  - Inorganic carcinogens.

Direct-Acting Carcinogens

- Direct-acting carcinogens or primary carcinogens interact directly with DNA.

- Direct-acting carcinogens are highly reactive (no activation is necessary). They may be chemically or enzymatically inactivated.

- Direct-acting carcinogens require only small amounts to reach target molecules.

- Direct-acting carcinogens are mostly mutagens.

- Examples include bis (chloromethyl) ether and ethylene dibromide.

Pro-Carcinogens

- Pro-carcinogens are also termed secondary carcinogens, and require conversion through metabolic activation by the host.

- Once activated, the activated form reacts rapidly with DNA and RNA.

- Pro-carcinogens are not all mutagens.

- Examples include methyl imine (metabolite hydroxymethyl amine is the carcinogen), vinyl chloride, benzo (a) pyrene, and 2-naphthylamine.

Inorganic Carcinogens

- Inorganic carcinogens are not directly genotoxic, but may lead to changes in DNA by selective alteration in fidelity of DNA replication.

- Examples include nickel and chromium.

EPIGENETIC CARCINOGENS

- Epigenetic carcinogens are those for which no evidence of genotoxicity has been found. Other possible mechanisms may involve chronic tissue injury, hormonal imbalance, immunologic effects, or promotional activity on cells that are either genetically altered or have been independently altered by genotoxic carcinogens.
- Solid-state carcinogens usually undergo long latent periods, with the exact mechanism unknown. Examples include asbestos.
- Hormone imbalances mainly alter endocrine system balance and differentiation, and often act as promoters. Examples include Estradiol and diethylstilbestrol (DES).
- Immunosuppressor carcinogens mainly stimulate virally-induced neoplasms.
- Co-carcinogens enhance effects when given at the same time. Examples include pyrene, catechol, ethanol, and sulfur dioxide (SO_2).
- Promoters enhance effects when given subsequently.
- Examples include phenol, anthralin, and saccharin. Benzo (a) pyrene is genotoxic (initiator); when later exposed to croton oil (promoter), skin cancer develops.

Mutagens and Teratogens

MUTAGENS

- Mutagenic agents include any number of chemical compounds able to induce mutations in DNA and in living cells.
- Mutations may or may not lead to cancer.
- An example is dimethyl sulfate, which reacts with the nitrogen atom of guanine, a constituent of RNA and DNA. The reaction affects the guanine molecule in such a way as to ultimately induce a mutation in DNA.

TERATOGENS

- Literal translation is "monster making."
- Refers to the production of abnormal offspring.
- Certain chemicals may produce malformations of a fetus without killing the fetus or damaging the mother.

- Damage to a fetus can occur at different times during pregnancy. Most likely to occur during early pregnancy (first trimester), when most women are not even aware they are pregnant.

Examples of Teratogens

- Agents that are presently known to be teratogens include:
  - Rubella (infection). Approximately 20% of women who contract rubella during the first trimester of pregnancy will give birth to a child with some deformity, ranging from cataracts and other eye anomalies, to deafness, cardiac effects, and mental retardation.
  - Thalidomide, ethylene oxide (chemicals). Used as a sedative hypnotic in the late 1950's in West Germany, England, and other countries. Relatively non-toxic to the user. Shortly after widespread use, there was an increase in infants with phocomelia (missing limbs). In most cases, thalidomide had been ingested between the third and eighth weeks of pregnancy.
  - Possibly steroids, ionizing radiation.

ACGIH Carcinogen Classification (Source: 1996-1997 TLV Booklet)

A1: CONFIRMED HUMAN CARCINOGEN

- A1 (confirmed human carcinogen) agents are carcinogenic to humans based on weight of evidence from epidemiologic studies of, or convincing clinical evidence in, exposed humans.

- Sample A1 carcinogens:
  - 4-aminodiphenyl.
  - Arsenic elemental and inorganic compounds (except arsine).
  - Asbestos.
  - Bis (chloromethyl) ether.
  - Chromite ore processing.
  - Chromium VI (water soluble and insoluble compounds NOC).
  - Coal tar pitch volatiles.
  - Beta-napthylamine.
  - Nickel sulfide roasting (fume and dust).
  - 4-Nitrodiphenyl.
  - Vinyl chloride.
  - Zinc chromate.

- Proposed additions to A1 list:
  - Benzene (change from A2).
  - Nickel (elemental and soluble compounds).
  - Oil mist (mineral, mildly refined).
  - Uranium (natural).
  - Wood dust (certain hard woods).

- Proposed deletions from A1 list: 4-Nitrodiphenyl (to A2).

A2: SUSPECTED HUMAN CARCINOGEN

- A2 (suspected human carcinogen) agents are carcinogenic in experimental animals at dose levels, by routes of administration (at sites, of histologic types, or by mechanisms that are considered relevant to worker exposure). Available epidemiologic evidence is either conflicting or insufficient.

- Sample A2 carcinogens:
  - Ethylene oxide.
  - Formaldehyde.
  - Methylene chloride.
  - Many others.

A3: ANIMAL CARCINOGEN

- A3 (animal carcinogen) agents are carcinogenic in experimental animals at relatively high doses, by routes of administration (at sites, of histologic types, or by mechanisms that are not considered relevant to worker exposure). Available epidemiologic evidence does not confirm an increased risk of cancer in exposed humans. Unlikely to cause cancer in humans except under uncommon or unlikely routes or levels of exposure.

- Sample A3 carcinogens:
  - Perchloroethylene.
  - Lead (elemental and inorganic compounds).
  - Many others.

A4: NOT CLASSIFIABLE AS A HUMAN CARCINOGEN

- A4 (not classifiable as a human carcinogen) are agents for which inadequate data on humans and/or animals is available.

- Sample A4 substances:
  - Ethylene glycol aerosol.
  - Chromium III compounds.
  - Inorganic mercury.

- There are many proposed A4 substances.

A5: NOT SUSPECTED AS A HUMAN CARCINOGEN

- A5 (not suspected as a human carcinogen) substances are not suspected as human carcinogens based on properly conducted epidemiologic studies in humans.

- Sample A5 Substance: Trichloroethylene.

7 Chemical Hazards

The skill of recognizing potential health hazards in the workplace is enhanced when industrial hygienists understand the characteristics of specific chemicals or hazardous materials and their toxicological effects.

Alcohols

Uses of Alcohols

- ❑ Uses are numerous. Generally used as solvents, co-solvents, and chemical intermediates.
- ❑ Alcohols with six to eleven carbon atoms are used in the manufacture of plasticizers. Alcohols with twelve to eighteen carbons are used in the production of detergents, surfactants, lubricants, and cosmetics.

Methanol

- ❑ Methanol (CH_3OH) is sometimes referred to as methyl alcohol.
- ❑ Metabolized to formaldehyde, formic acid, then formate. Formate causes damage to either/or both the retina and optic nerve, resulting in visual impairments and possibly blindness.
- ❑ Acute poisoning symptoms follow a pattern: a narcotic effect (CNS depression), followed by an asymptomatic interval (12-24 hours), then headaches, nausea, dizziness, vomiting, severe abdominal pain, and difficulty in breathing.
- ❑ Symptoms also involve blurred vision, photophobia, and pain in the eyes. Visual disturbances may either recede or develop into total blindness. Causes blindness if ingested.

Ethanol

- ❑ Ethanol (CH_3CH_2OH) is sometimes referred to as ethyl alcohol.
- ❑ Ethanol is irritating to eyes and nose, and causes headaches.
- ❑ Symptoms of poisoning include: fetal alcohol syndrome (due to alcoholic beverages); decreased growth; joint, limb and cardiac anomalies; and behavioral and cognitive impairment. No reports of fetal alcohol syndrome have occurred from occupational exposure.

Other Alcohols

- ❑ 1-Propanol ($CH_3CH_2CH_2OH$):
  - ❑ Not very irritating.
  - ❑ Low acute and chronic toxicity.
  - ❑ May produce sensory irritation at very high concentrations.

- Isopropanol ($CH_3CHOHCH_3$):
  - Low acute toxicity.
  - Irritating to eyes but not skin.
  - Prolonged exposure may produce CNS depression and narcosis.

Aldehydes

Aldehyde Uses

- Aldehydes are intermediates in the synthesis of chemicals important in the construction, agricultural, garment, pharmaceutical, and rubber industries. Higher molecular weight aldehydes are used to contribute characteristic flavors and odors to perfumes and oils.

- Aldehydes can exist as gases (e.g., formaldehyde), liquids, or even solids (heterocyclic aldehydes).

Aldehyde Examples

- The interest in the toxicity of aldehydes has focused on three specific compounds:
  - Formaldehyde (CH_2O), used as a fumigant for grains and particle board glue.
  - Acetaldehyde (CH_3CHO), which also presents an explosive hazard.
  - Acrolein ($CH_2{=}CHCHO$), found in tear gas, weed killer, and a byproduct of burning oil.

- The toxicity of these materials in demonstrated in the table below.

	Oral LD50 Rats	TLV (ppm)
Formaldehyde	0.5 g/kg (very to moderate)	C 0.3
Acetaldehyde	1.9 g/kg (moderate)	C 25
Acrolein	0.046 g/kg (extreme to very)	0.1, C 0.3
Note: LC50's are probably more relevant.		

Aldehyde's Toxic Effects

- Aldehyde's toxic effects include:
  - Irritation of mucous membranes.
  - Sensitization or allergic reaction (observed with formaldehyde).
  - Anesthetic properties.

- Laboratory animals exposed to aldehyde vapors show damage to the respiratory tract and pulmonary edema. This is much more pronounced for the very reactive aldehydes (acrolein) and often includes deep lung damage.

- Carcinogenicity is still being questioned for formaldehyde (A2), acetaldehyde (A3), and acrolein (no designation).

- Formaldehyde has been shown to cause tumors in the respiratory passages of rats and possibly mortuary workers.
- The irritant properties generally limit substantial worker exposure.

Alicyclic Hydrocarbons

Cycloalkanes

- Cycloalkanes are sometimes referred to as cycloparaffins.
- Lower cycloalkanes are gases, and have been used as anesthetics (e.g. cyclopropane).
- For those molecules with more than six carbon atoms, the margin of safety between anesthesia and death (due to cardiac sensitization and CNS depression) is narrow (small therapeutic index—TI).
- For those molecules with five or more carbon atoms, the members are liquids and may cause CNS depression.
- Generally, cycloalkanes are CNS depressants with relatively low acute and chronic toxicity (examples: cyclopentane, cyclohexane).

Cycloalkenes

- Cycloalkenes are sometimes referred to as cycloolefins. They are more highly reactive than cycloalkanes.
- Cycloalkenes act as irritants and some have sensitizing properties.
- Examples:
  - Limonene occurs in oil of many plants, has a highly fragrant citrus odor, is a dermal irritant, and may be a skin sensitizer.
  - Turpentine is a moderate health hazard that is readily absorbed through the skin or respiratory tract. It irritates skin, eyes, nose, and mucous membranes. Major systemic effects include kidney and bladder injury.

Aliphatic and Alicyclic Amines

Aliphatic and Alicyclic Amines

- Aliphatic amines (e.g., methylamine-CH_3NH_2) are highly alkaline and tend to be fat-soluble.
- Exposure may result in skin and mucous membrane irritation. Corrosive burns and allergic sensitization may occur.

- Most have an unpleasant, fish-like odor. In high concentrations, they have an ammonia-like odor. Olfactory fatigue can occur rapidly.

- Primary toxic reactions are irritation and sensitizing effects. Vapors can cause eye irritation, with lacrimation, conjunctivitis, and corneal edema, which result in seeing "halos" around light. Direct contact can cause permanent eye damage and skin burns.

- Systemic symptoms from inhalation include headache, nausea, faintness, and anxiety.

- Sample compounds include:
  - Methylamine.
  - Ethylamine.
  - Ethylenediamine.
  - Cyclohexylamine.

- Ethylenediaminetetraacetic acid (EDTA) :
  - Has been used in the form of a calcium disodium salt (calcium disodium edetate), and as a chelating agent for the removal of toxic heavy metals.
  - Used as beginning materials for chemical synthesis, intermediates, and as solvents.

Aliphatic Hydrocarbons

Alkanes

- Alkanes (paraffins, saturated hydrocarbons) include methane, ethane, pentane, etc.

- Toxicity of alkanes is generally related to vapor pressure, viscosity, surface tension, and lipid solubility.

- Can cause nausea, vomiting, abdominal pain, and diarrhea when ingested. Generally considered not very toxic for single doses below the lower flammability limit. Lower alkanes (C_1-C_3) are considered simple asphyxiants.

- Serious toxic effects of alkanes include:
  - Asphyxia (displacing oxygen).
  - Chemical pneumonitis.
  - Axonal neuropathy (n-hexane).
  - Cancer for butadiene (an alkene).

- Dermatitis.
- CNS depression.
- Anesthesia.
- Cardiac sensitization.

Normal Hexane

- Normal hexane (n-hexane) may be the most toxic alkane.
- If ingested, normal hexane causes nausea, vertigo, bronchial and intestinal irritation, and CNS depression.
- Has also been shown to induce axonapathy through a metabolite, 2,5-hexanedione (actually, the pyrole formed from 2,5-hexanedione is the ultimate causative agent).
- In lab animals exposed to 10,000 ppm, six hours/day, and five days/week for 13 weeks, there was little axonapathy.
- Continuous exposure to as little as 250 ppm for four months produced severe axonapathy.
- Formation of metabolite lags eight to 10 hours after exposure. With eight hours or less of exposure, the metabolite 2,5-hexanedione can be excreted in urine. BEI is less than 5mg/g creatinine at end of shift. With longer-term exposure, the metabolite builds to a steady-state level in the body with resultant neurotoxicity.

Alkenes

- Alkenes (olefins) are generally slightly more toxic than alkanes.
- No alkenes are known to produce axonapathy. Some have anesthetic properties: ethylene, propylene, and pentene. Some are simple asphyxiants, such as ethylene.
- Longer chain compounds are mucous membrane irritants and have cardiac effects.
- Example: Butadiene (CH_2=CHCH=CH_2) has caused thyroid tumors in rats and mice. TLV class A2.

Alkynes

- The lower members of the alkynes class are generally anesthetics.
- Toxic effects include:
  - Central nervous system depression.
  - Pulmonary irritation and edema at high concentrations.
  - Ability to act as a simple asphyxiant.

Aliphatic Nitrates

Uses and Hazards of Aliphatic Nitrates

- The nitrate group has the structure C-O-NO_2.

- Used as explosives and blasting powders, particularly nitroglycerin (NG) and pentaerythritol tetranitrate (PETN). Methyl, ethyl, propyl, and isopropyl nitrates have also been used as rocket propellants and jet fuels.

- Primary effects of aliphatic nitrates are:
  - Dilation of blood vessels (lowering the blood pressure).
  - Methemoglobin formation (chemical oxidation of heme iron in hemoglobin, resulting in a decreased ability of hemoglobin to carry either oxygen or carbon dioxide).
  - Blood vessel dilation also leads to headaches.
  - Severe headaches associated with nitroglycerin exposure resulted in the use of the term "dynamite head" and "powder headache."
  - Nitroglycerin is also associated with "Monday morning angina," a cardiovascular condition resulting from nitrate withdrawal (one to three days after last exposure).
  - Nitroglycerin or trinitroglycerol or glycerol trinitrate formula: $C_3H_2(ONO_2)_3$.

- Alkyl nitrites (methyl nitrite, ethyl nitrite, amyl nitrite, etc.) have many of the same pharmacological and toxicological effects, namely vasodilation, tachycardia, and methemoglobin production.

Alkaline Materials

Alkaline Materials

- Primary health hazards of alkaline materials include irritation and corrosion of tissues with direct contact. Most exposures are the result of spills and splashes.

- Examples include:
  - Ammonia (NH_3) and ammonium hydroxide.
  - Calcium hydroxide.
  - Calcium oxide.
  - Potassium hydroxide.
  - Sodium hydroxide.

- Potassium and sodium—reactive metals—can be either corrosive in direct contact with the skin, or from solutions they form with water.

Aromatic Hydrocarbons

Uses and Hazards of Aromatic Hydrocarbons

- Benzene's derivatives and homologues are used as solvents and are often components of commercial and consumer products.

- Major areas of interest for this class of compounds are:
  - Hematopoietic effects of benzene (aplastic anemia).
  - Benzene induced leukemia.
  - Cerebellar lesions and loss of CNS integrative functions in "glue sniffers" as a result of toluene.
  - Carcinogenicity of styrene and PAHs.

- Two forms of chronic toxicity include:
  - Cancer (acute myelogenous leukemia in adults).
  - Injury to the hematopoietic system (aplastic anemia).

Health Effects of Aromatic Hydrocarbons

- Aromatics are generally skin irritants.

- Aromatics are more toxic than aliphatics with the same number of carbon atoms. They may result in dermatitis by dehydrating and defatting the skin. They are known for their central nervous system effects.

- Effects are also seen in liver, kidneys, spleen, bladder, thymus, brain, and spinal cord in animal tests. Aromatics have a special affinity for nerve tissue.

Benzene

- Benzene is readily absorbed through the skin, lungs, and gastro-intestinal tract (GIT).

- Hematopoietic effects of benzene include aplastic anemia and benzene induced leukemia.

- Metabolites of benzene are thought to be the causative agent in chronic toxicity (open ringed metabolites).

- Widely distributed in the body, with affinity for fatty tissues, benzene also tends to deposit in bone marrow, the lungs and the kidneys.

- Primarily metabolized in the liver by cytochrome P-450. Also metabolized in other tissues.

- Competitive inhibitors of its metabolism (e.g., toluene) may limit the toxic effect.

- Measuring total phenol in urine at the end of the shift of less than 50-mg/g creatinine requires biological monitoring.

Toluene

- Toluene is sometimes referred to as methylbenzene. Its toxicity is similar to benzene, but it is not hemotoxic or carcinogenic.
- Toluene depresses the central nervous system. It is a skin and mucous membrane irritant. Drying and defatting of skin causes dermatitis.
- Toluene is also a cardiac sensitizer, and can cause hepatomegaly and nephrotoxicity.

Xylenes

- Xylenes include ortho-, meta-, and para-forms and is sometimes referred to as dimethyl benzene.
- Xylene is an irritant, depresses the central nervous system, and is a cardiac sensitizer.

Styrene

- Styrene is sometimes referred to as cinnamene, or vinylbenzene. It is highly reactive and readily polymerizes. Violent explosions sometimes accompany it.
- Styrene is used in plastics and resins and often as a chemical intermediate.
- Styrene is irritating to the eyes, skin, mucous membranes, and respiratory system. It has been shown to be carcinogenic in rodent inhalation studies. It has no TLV classification.

Other Aromatic Hydrocarbons

- Naphthalene (moth flake) is:
  - A white solid with mothball or tar odor (old mothballs).
  - Toxic—responses to large exposures involves hemolytic anemia.
  - Irritating to eyes.
  - Retinotoxicity has been reported.
- Ethylbenzene is irritating to the skin and eyes.
- Isopropyl benzene or cumene is:
  - An eye and skin irritant.
  - A CNS depressant, causing dizziness, incoordination, and unconsciousness.

Polycyclic Aromatic Hydrocarbons

- Polycyclic aromatic hydrocarbons (PAHs) are sometimes referred to as polynuclear aromatic hydrocarbons (PNAs), and are significant contributors to smog formation.
- Several of them are carcinogenic or suspected carcinogens (e.g., benzo-a-pyrene, dibenzo (a,h) anthracene, benzo(a)anthracene, benzofluoroanthracene, and chrysene).
- Acute toxicity is generally pretty low, but chronic exposures are associated with skin cancer.

Hexachlorobenzene—A Halogenated Benzene

- Hexachlorobenzene (HCB) (C_6Cl_6) has been shown to be immunotoxic (increases susceptibility to infectious agents) in experimental animals.
- An epidemic of HCB toxicity occurred in Turkey between 1955 and 1959, due to its use as a fungicide on wheat.
- People have suffered from vesicular diseases after exposure, as well as hepatomegaly and porphyria (clinical disorders that are either genetically acquired or chemically induced, involving abnormal heme synthesis that ultimately leads to heme depletion), resulting in *porphyria cutanea tarda*.
- Other adverse health effects of hexachlorobenzene include cutaneous, arthritic, and neurologic effects.
- Many toxic effects have lasted 20 years after exposure.
- Increased liver cancer seen in animal studies, but not during toxic episode in Turkey.

Other Halogenated Benzenes

- Other halogenated benzenes include Chlorobenzene, ortho-Dichlorobenzene, and para-Dichlorobenzene.
- Chlorobenzene (C_6H_5Cl):
  - Solvent used for pesticides and degreasing.
  - A CNS depressant.
  - Lesions of liver and kidney also seen.
- o-Dichlorobenzene (o-DCB) produces liver and kidney damage.
- p-Dichlorobenzene (p-DCB) causes eye and nose irritation, and lung, liver and kidney damage.

Aromatic Nitro and Amino Compounds

Uses and Hazards of Aromatic Nitro and Amino Compounds

- Uses include production of polymers, dyestuffs, pesticides, pharmaceuticals, rubber products, explosives, adhesives, and paints.
- Many of these compounds are methemoglobin formers, some are sensitizers, and others are known animal and/or human carcinogens. Some may cause liver, kidney, spleen, and testicular damage.
- Generally, liquid and vapor forms are readily absorbed through the intact skin.
- Methemoglobin (or ferrihemoglobin) is unable to transport oxygen, and symptoms of oxygen deficiency may begin to appear. Cyanosis (bluish color beginning with the lips and ears) may occur in humans at level of 15–25% methemoglobin.
- Information suggests that aromatic nitro and amino compounds are only methemoglobin inducers after being metabolized in the body.
- The carcinogenicity caused by certain aromatic nitro and amino compounds is also thought to be the result of metabolism to an electrophilic species that is capable of alkylating genetic material.

Examples of Aromatic Nitro and Amino Compounds

- Examples of aromatic nitro and amino compounds include:
    - Aniline.
    - 4,4'-methylene bis (2-chloroaniline) (MOCA).
    - Diphenylamine.
    - Nitrobenzene.
    - Benzidine.
    - Toluidines.
- Three aromatic amines that are generally considered to be carcinogens (all have been shown to be associated with bladder cancer) are:
    - 4-Aminodiphenyl.
    - Benzidine.
    - B-Napthylamine.
- Nitrobenzene and 1,3-dinitrobenzene are related to reproductive effects in laboratory animals through damage to the testes.
- 2,4,6-trinitrotoluene (TNT) is irritating to eyes, nose, throat, and skin. Headaches, gastritis, toxic hepatitis, bone marrow depression, and cataracts have been associated with exposure.

Chlorinated Hydrocarbons

Chlorinated Hydrocarbon Insecticides

- Chlorinated hydrocarbon insecticides were used extensively in the US and other western countries and continue to be used in third world countries.
- Examples include Aldrin, Dieldrin, Chlordane, Heptachlor, DDT, Endrin, Kepone, Lindane, Mirex, and Toxaphene.
- DDT (dichlorodiphenyltrichloroethane) is credited as the primary compound that brought epidemics of malaria, typhus, and plague to a complete stop.
- DDT was introduced in 1943. It is still used today to combat malarial mosquitoes in tropical regions, where substitution with other insecticides, such as Malathion, would be very expensive.
- The persistence of pesticides in the environment provided not only an immediate impact, but also a prolonged insecticide presence. This is now considered a negative aspect due to its delayed adverse effects on non-target populations of insects and birds.
- An increased cancer risk for humans has been alleged, but not proven.
- Use of these compounds in the US has been virtually eliminated (although lindane has some limited use).
- Primary acute toxicity in humans and animals is neurological hyperactivity, such as CNS stimulation and tremors or convulsions.
- Studies in mice have shown benign and malignant liver tumors. This effect has not been shown in humans.

Cyanides and Nitriles

Cyanides

- Cyanide readily forms stable complexes with a number of biologically active metal ions, particularly with Fe^{+3} in cytochrome oxidase. Since cytochrome oxidase plays a central role in the utilization of oxygen in most cells, its inhibition leads to a loss of cellular functions and to cell death.

- Most specific pathological findings in acute cases are the bright red color of venous blood, which is an indicator of inability of tissue cells to use oxygen. Cyanide does not bind with hemoglobin in blood, but it does bind somewhat with methemoglobin (Fe^{+3} normally accounts for about 2% of hemoglobin). This is also a classic toxicologic response to hydrogen cyanide (HCN), also known as prussic acid.

- Antidote for poisoning involves two steps:
  1. Methemoglobin forming agent (levels of up to about 20% methemoglobin are not usually harmful, this methemoglobin can then transport some of the cyanide).
  2. Thiosulfate (increases formation of less toxic thiocyanate).

- These therapies are generally combined with artificial respiration and amyl nitrite therapy.

- Other forms of cyanide include sodium cyanide, potassium cyanide, and cyanamide.

Nitriles

- Some nitriles display toxic effects similar to cyanide. Acrylonitrile's (vinyl cyanide) symptoms of toxicity resemble cyanide poisoning.

- Not all nitriles dissociate to produce cyanide. They are sometimes referred to as organic cyanides, but this is not appropriate since they do not act like cyanide. They often have other toxic effects, however.

- A suspected human carcinogen (A2), nitriles cause astrocytomas and squamous cell carcinoma of the gastro-intestinal tract (GIT) in rats.

Hydrazine

- Hydrazine (H_2N-NH_2) is used as rocket fuel, as a reducing agent, and as an oxygen scavenger.

- The skin, gastro-intestinal tract, and lungs rapidly and thoroughly absorb hydrazine.

- Acute toxicity is accompanied by liver damage, and red cell destruction (anemia), as well as pathological changes in lung, liver, and kidneys. Severe skin irritation, burns, corrosion, and sensitization may also occur.

- Hydrazine has been shown to be carcinogenic (A3) in laboratory animals, causing lung and liver cancer and myeloid leukemia.

Phosgene

- Phosgene ($COCl_2$) is a gas at ambient temperature and pressure. It has a mild odor and weak acute irritant properties, offering only a limited warning of its potential toxicity.
- Phosgene was used as a chemical warfare agent in World War I.
- Progressively causes tracheitis, bronchitis, and pulmonary edema.

Epoxy Compounds

Epoxy Compounds—Types and Uses

- Epoxy compounds are any compounds containing one or more oxirane rings. Oxirane rings are oxygen atoms linked to two adjacent carbon atoms, forming a highly reactive structure.
- Industrial uses include uncured epoxy resins that require curing agents to form the desired hard polymers.
- Uncured resins are the main toxicological interest, since cured resin should not have any unreacted epoxide groups remaining.
- Epoxy compounds include ethylene oxide, propylene oxide, and epichlorohydrin. Ethylene oxide, propylene oxide, and butylene oxide are all highly flammable.
- Ethylene oxide (EO or EtO) (C_2H_4O):
  - Also known as 1,2-epoxyethane, oxirane, and dimethylene oxide.
  - EtO is reactive and highly irritating to exposed tissues.
  - EtO is a direct (not requiring metabolism) mutagen in many test systems (bacteria, fungi, yeasts, mammalian cells).
- Epichlorohydrin (ECH) (CH_2OCHCH_2Cl):
  - Flammable, reactive liquid.
  - Sweet, pungent or chloroform-like odor.
  - Results in a dermatitis reaction, with blisters in severe cases.
  - Can penetrate leather shoes and rubber gloves.

Health Effects of Epoxy Compounds

- Adverse health effects are dependent on the reactivity of the epoxide, molecular weight, and solubility. These include:
  - Dermatitis (including sensitization dermatitis).
  - Eye irritation, pulmonary irritation, and gastric irritation.
  - Some suggestion of systemic effects on the liver and kidney.

- Normally, the irritant properties limit exposure and therefore limit potential for systemic effects.
- Many epoxide compounds have been found to be carcinogenic in rodents.
- Human epidemiology studies have been unclear.
- Rodent cancers generally occur at the tissue of first contact (includes propylene oxide, epichlorohydrin, and others).
- Few epoxides (e.g., ethylene oxide) have produced tumors at sites other than first contact.
- Ethylene oxide has been implicated as a developmental toxin (teratogen) based on an IV study in mice.

Esters

Uses and Health Effects of Esters

- Uses include additives in automotive and aircraft industries, flavor and fragrance additive in food and cosmetics, use in lacquers and general solvents, plastics, and resins.
- Example compounds include methyl formate, ethyl formate, methyl acetate, ethyl acetate, and methyl acrylate.
- Highly irritating, may also cause dermal sensitization.
- General toxic responses include primary irritation to eyes, skin, respiratory and gastro-intestinal tracts and CNS depression. Also can have impact on GI, hepatic, and renal systems. Known to cause abnormal cardiac rhythm.
- Other effects include hypersensitization, reproductive effects (including teratogenesis), and carcinogenesis. Most commonly observed effects are irritant and anesthetic effects.

Ethers

Types and Uses of Ethers

- Used in a variety of industries, including rubber, plastics, paints, medicine, and dentistry. More volatile ethers have been used as liquid refrigerants, fuel additives, and rocket propellants.

- The ethers that have occupational exposure standards are:
  - Diethyl ether.
  - Diisopropyl ether.
  - Phenyl ether.
  - Diphenyl etherdiphenyl mixture.
  - Chloromethyl methyl ether (ACGIH A2 carcinogen, although no exposure limit).
  - Bis (chloromethyl) ether (ACGIH A1 carcinogen, although no exposure limit).
  - Dichloroethyl ether.
  - Chlorinated diphenyl oxides.

- Ethers exposed to oxygen in air may form peroxides, presenting an explosion hazard.

Examples of Ethers and Health Effects

- Diethyl ether:
  - Commonly referred to as ether (the medical anesthetic).
  - Absorbed quickly through the lungs and eliminated quickly by the lungs.
  - Concentrations of 10–15% in air are needed for anesthesia induction.
  - Anesthesia can be maintained by concentrations of about 5%. Fatalities may result at concentrations above 10%.
  - Ether is irritating to mucous membranes.

- Methyl t-butyl ether (MTBE), a gasoline additive.

- Chloromethyl methyl ether (CMME). Oat cell lung cancer has been identified in workers exposed to CMME.

- Bis (chloromethyl) ether (BCME) has been shown to be both mutagenic and carcinogenic. In six studies of exposed workers, lung tumor incidence was increased (oat cell carcinomas).

Organic Peroxides

- Organic peroxides are highly reactive molecules due to the oxygen-oxygen linkage.

- Under activation conditions (e.g., heating), the oxygen-oxygen bond is cleaved to form highly reactive free radicals.

- These free radicals can be used to initiate curing or polymerization.

- If free radicals form during storage, decomposition may result, leading to the release of considerable heat and energy.
- Because of physical hazards, only acute toxicity testing has been performed on organic peroxides.
- Most organic peroxides are irritating to the skin and mucous membranes.
- Example: Dibenzoyl peroxide (benzoyl peroxide, BPO) (C_6H_5-C-O-O-C-C_6H_5) has been shown to act as a tumor promoter in mice and hamsters when repeatedly applied to skin. Uses include bleaching of flour, additive in self-curing of plastics, and topical preparations to treat acne (antibacterial action).

Fluorine Compounds

Organic Compounds Containing Fluorine

- When chlorofluorocarbons (CFCs) were introduced in the 1940's, they were considered inert refrigerants compared to sulfur dioxide, ammonia, carbon tetrachloride, and chloroform, which were already in use.
- The use and abuse of certain CFCs in aerosols led to accidental deaths by cardiac arrest (by sudden sniffing death syndrome). Prototype CFC is trichlorofluoromethane (CFC_{11}).
- Trichlorofluoromethane is the most widely used propellant, and has been proven to be cardiotoxic (causing cardiac arrhythmia and tachycardia) at high exposures.
- Types of organic compounds containing fluorine include:
  - Chlorofluorocarbons (CFCs)—e.g., trichlorofluoromethane (CCl_3F).
  - Hydrogenated CFC (HCFC)—e.g., dichlorofluoromethane ($CHCl_2F$).
  - Hydrogenated fluorocarbon—e.g., difluoroethane.
  - Fluorocarbon—non-chlorinated and non-hydrogenated.
- It is generally agreed that CFCs photodissociate in the stratosphere, producing chlorine atoms that lead to the destruction of stratospheric ozone.
- Reduction of stratospheric ozone allows more UV light to reach earth and will likely lead to increases in:
  - Cataracts.
  - Immunologic diseases.

- Malignant melanomas (a serious form of skin cancer causing death).
- Basal squamous cell carcinoma of the skin (less serious, but more prevalent).

- Trichlorofluoromethane and dichlorofluoromethane can lead to stratospheric ozone depletion, CFCs being more potent ozone depleters.

- Fluorine containing hydrocarbons are less toxic than the chlorinated hydrocarbons (e.g., trichlorofluoromethane is less toxic than CFC_{14}).

- The general rule of thumb is "the more chlorinated the more toxic." CFC vapors are four to five times heavier than air, therefore they accumulate in low-lying areas.

- In addition to refrigerants, other uses include solvents, foam blowing, fire extinguishing, and aerosol propellants.

Glycol Ethers

Health Hazards of Glycol Ethers

- Methyl and ethyl ethers of ethylene glycol have been shown to be reproductive and developmental toxicants in animal studies.

- Propyl and butyl ethers of ethylene glycol have not been shown to be reproductive or developmental toxicants in animal studies.

Ethylene Glycol Monomethyl Ether

- Ethylene glycol monomethyl ether is known as EGMME.

- EGMME is mildly irritating to skin and eyes. It is absorbed through the skin but has low acute toxicity by this route.

- EGMME is more toxic when inhaled. Toxic actions are on the brain, blood, kidneys, thymus, and testes.

- Is embryotoxic and teratogenic (e.g., encephalopathy) in all animal studies.

Ethylene Glycol Monoethyl Ether

- Ethylene glycol monoethyl ether (EGMEE) has low acute oral toxicity.

- Other health effects include:
  - Slight irritation to eyes and mucous membranes.
  - Absorption through skin, though not very toxic by this route.
  - Somewhat more toxic when inhaled.
  - Sites of toxic action are blood, liver, kidneys, and testes.

- Considered a developmental toxic and teratogen based on animal studies.

Ethylene Glycol Mono-n-Propyl Ether

- Ethylene glycol mono-n-propyl ether is irritating to eyes.
- Other health effects include:
  - Slight irritation to skin.
  - Skin sensitization.
  - Primary site of toxicity is red blood cells.
  - Secondary effects in liver, spleen, and kidney (hemoglobin present in urine).
- Has not been shown to have teratogenic or developmental effects.

Glycols

Uses and Health Effects of Glycols

- Glycols are hydrocarbons with two hydroxyl groups attached to separate carbons in an aliphatic chain.
- Glycols are used as antifreeze agents, solvent carriers, and chemical intermediates.
- Since these compounds have low vapor pressures, inhalation of vapors or aerosols is not a concern. Glycols have little or no odor. Glycols can be irritating at high levels, and there is a potential for central nervous system depression.

Ethylene Glycol

- Ethylene glycol is a commonly used anti-freeze.
- It presents negligible health hazards in industrial handling and has low acute toxicity.
- Ethylene glycol is not significantly irritating to eyes or skin, and is not readily absorbed through skin.
- It has low vapor pressure at room temperatures.

Propylene Glycol

- Propylene glycol is also used in anti-freeze. Other uses include brake fluid, hydraulic fluid, and the manufacturing of resins.
- Propylene glycol has a very low toxicity. Since 1942, it has been considered an acceptable ingredient for pharmaceutical products.

Halogenated Aliphatic Hydrocarbons

Uses and Health Effects of Halogenated Aliphatic Hydrocarbons

- As a class of compounds, halogenated aliphatic hydrocarbons are widely used in industry. Often used as solvents for dry cleaning of clothing and cleaning of electrical equipment and metals.
- Although they primarily enter the body through inhalation, they can be absorbed through the skin.
- The health effects are generally narcotic, affecting the central nervous system.

Methylene Chloride

- Methylene chloride (CH_2Cl_2) is used as blowing agent for foams, solvents, etc. Methylene chloride is a very volatile substance.
- Methylene chloride's primary toxic effect is narcosis. May cause "drunkenness" leading to inept operation of equipment. Symptoms include dizziness, nausea, tingling or numbness of extremities, stupor, and lethargy.
- Methylene chloride is metabolized to carbon monoxide.
- There is some evidence of carcinogenicity in animals. No carcinogenicity in humans has been reported (A2 classification).

Chloroform

- Chloroform is sometimes referred to as trichloromethane ($CHCl_3$).
- Chloroform was used widely as anesthetic, but due to delayed liver injury and (lesser extent) cardiac sensitization, this use is now obsolete. Some use continues as a solvent, primarily a chemical intermediate.
- Acute exposure results in central nervous system depression, narcosis, and death. Concentrations below anesthetic level result in inebriation and excitation. High concentrations may result in cardiac sensitization. Liver injury is common from chronic exposure.

Carbon Tetrachloride

- Carbon tetrachloride (CCl_4) is used little today due to hepatic and renal toxicity and the availability of less hazardous solvents.
- Acute exposures result in CNS depression and possibly cardiac sensitization. Liver and kidney injury may occur after single, acute exposure, but is more likely to occur from chronic exposure. The liver is generally affected before the kidneys.

- Toxic effects are the result of a metabolite resulting in many strain and species differences in toxicity.
- Carcinogenic effect (A3) may be the result of organ injury.

Methyl Chloroform

- Methyl chloroform is sometimes referred to as 1,1,1-trichloroethane.
- Its major use is as a solvent.
- High volatility with careless use. Results in death from gross overexposure (usually in confined spaces) or due to deliberate inhalation.
- Future use of methyl chloroform is questionable due to ozone depletion.
- Methyl chloroform is the least toxic of the chlorinated solvents, although it can cause CNS depression and cardiac sensitization at high concentrations.

Vinyl Chloride

- Vinyl chloride is sometimes referred to as monochloroethylene, or chloroethene ($CH_2{=}CHCl$).
- Vinyl chloride gas is used as a monomer in plastic manufacture.
- Original major concerns were fire and explosion hazards.
- In 1974, three cases of angiosarcoma of liver were reported in grossly exposed workers. Vinyl chloride also has a hepatotoxic action.
- Latency of development of liver cancer from year of first exposure averages about 24 years.

Polyvinyl Chloride

- Polyvinyl chloride has historical significance as a causative agent in "meat wrappers' asthma." It was induced in meat packers when vapors were emitted from plastic that was heated and wrapped around meat.

Trichloro-ethylene

- Trichloroethylene, or 1,1,2-trichloroethylene, was a widely used solvent, particularly for metal degreasing and extraction processes.
- Acute responses of trichloroethylene include:
  - CNS depression.
  - Visual disturbances.
  - Mental confusion.
  - Fatigue.
  - Sometimes nausea and vomiting.

- "Degreaser's flush" (extreme redness of face and arms) may occur if alcohol is consumed shortly before or after exposure.

Perchloroethylene

- Perchloroethylene is sometimes referred to as tetrachloroethylene, or Perc ($CCl_2{=}CCl_2$).
- Perchloroethylene is used for dry cleaning and degreasing. Today it is a common groundwater contaminant from historical use as an industrial solvent.
- Liver and CNS effects have been reported following exposure.
- Massive oral doses in mice produce liver cancer, but not in rats. Kidney tumors have occurred in male rats.
- Limited epidemiologic evidence has been inconsistent or negative regarding cancer (TLV = A3).

Halogenated Cyclic Hydrocarbons

Polychlorinated Biphenyls

- A common halogenated cyclic hydrocarbon is the class of compounds referred to as polychlorinated biphenyls (PCBs).
- Trade names for this class of compounds include Aroclor 1242 and Aroclor 1260.
- Are known environmental hazards due to their stability in environment. Tend to bioaccumulate. Levels have been measured in fish, birds (especially eagles), human breast milk, human hair, etc.
- PCBs are not known to cause cancer (no TLV carcinogen classification exists) and have been known to cause hepatomegaly, peripheral neuropathy, and CNS effects. Usually absorbed through the skin (skin notation on TLV).
- First toxicity was recognized in Japan ("Yusho incident") where rice oil became contaminated with PCBs (and likely other agents such as dioxins and furans).
- Main toxic reactions of PCBs include chloracne (acneform dermatitis), liver impairment, and neuronal effects. Other effects include vague aches and pains and severe headaches. PCBs have an affinity for fatty tissue and can be transferred to mother's milk.

Halogenated Dibenzo-p-dioxins and Dibenzofurans

- Another class of halogenated cyclic hydrocarbons includes halogenated Dibenzo-p-dioxins and Dibenzofurans.
- As a group they are considered to be highly toxic.
- There are no commercial uses of these compounds.
- Potential contaminants in the manufacture of different products (e.g., pulp and paper manufacturing).
- Are the by-product of certain fires (e.g., fires involving PCBs).
- Generally, furans are considered slightly less toxic than their corresponding dioxin.
- 2,3,7,8 tetrachloro-p-dioxin (TCDD)—often referred to as dioxin—has the following qualities:
  - Most potent carcinogen tested in lab animals (liver tumors, thyroid gland, lungs, oral and nasal epithelia).
  - Toxic responses include immunotoxicity, teratogenicity, weight loss, hepatotoxicity, and porphyria.
  - Exposure will also result in chloracne, hyperpigmentation, muscle aches and pains, etc.
- Other dioxins and furans include:
  - 1,2,3,7,8 Pentachloro-p-dioxin (or furan).
  - 1,2,3,6,7,8 hexachloro.
  - Heptachloro.
  - Octachloro (lower toxicity).

Ketones

Uses and Health Effects of Ketones

- Ketones are organic compounds chemically structured with a carbonyl group (C=O) attached to two carbon atoms.
- Commonly used as solvents, extractants, chemical intermediates, and, to a lesser degree, as flavor and fragrance ingredients.
- Primary routes of exposure: inhalation and skin contact. Due to high vapor pressure at room temperature, inhalation is most likely route of exposure.

- Classic symptoms of overexposure include, in order of progression: irritation of eyes, nose, throat; headache; nausea; vertigo; in-coordination; CNS depression; narcosis; and cardiorespiratory failure. Overall relative toxicity of most ketones is low. Effects of chronic exposure are less understood than the acute.

- Ketone exposure may alter toxicity of other chemicals, including other ketones that are metabolized by Cytochrome P-450 enzymes.

- Certain ketones are neurotoxins (e.g., 2,5 hexanedione, methyl, n-butyl ketone, and ethyl n-butyl ketone). Neurotoxicity is result of common metabolite. The effect is characterized by nerve degeneration showing large, multifocal axonal swellings (referred to as "giant axonal" neuropathy). Myelin damage may also occur. Symptoms include bilaterally symmetrical parasthesia (pins and needles feeling), and muscle weakness in legs and arms. Ketones otherwise are relatively non-toxic.

- Most commonly used ketones (i.e., acetone, methyl ethyl ketone (MEK), methyl isobutyl ketone (MIBK), and cyclohexanone) appear to be the least toxic.

- MEK and MIBK have been shown to potentiate hepatotoxic effects of carbon tetrachloride and chloroform.

Acetone

- Acetone synonyms are 2-propanone and dimethyl ketone (C_3H_6O). One of the least toxic solvents in commercial use today, it has a sweet, fruity odor at low concentrations, and is highly flammable and volatile.

- Acetone is absorbed through the lungs and quickly distributed throughout the body.

- Normal by-product of mammalian metabolism. Body levels vary based on nutrition, starvation, alcoholism, diabetes, etc.

- Disulfiram (used to treat alcoholics) results in increase in level of acetone in circulating blood as well as expired air.

- Isopropanol is also metabolized to acetone, leading to increased levels of acetone in the blood.

- Acute, high level exposure results in irritation of mucous membranes and CNS depression, ranging from lightheadedness to narcosis.

- TLV = 750 ppm. Chronic toxicity is minimal.

Metals

Introduction

- As a group, metals have some similar toxicological properties. Many tend to accumulate in kidneys and bone.
- Important considerations of a metal are the route of exposure and whether it is an essential metal. Many have long half-lives.
- Concentration in urine is a good indicator of exposure and is a good marker of total body burden, even after exposure has stopped.

Aluminum

- Aluminum is used structurally in the building industry. It is also used in consumer durables and packaging.
- Aluminum use is often in the form of alloys, which gives aluminum more mechanical strength. Alloys often contain other metals, such as magnesium, copper, zinc, nickel, and silicon.
- Aluminum has limited absorption, rapid urinary excretion, and slow tissue uptake. Chronic exposure can result in the buildup of a body burden of aluminum.
- Shaver's disease was first described in 1947 for a group of furnace workers processing bauxite (aluminum oxide). Actual exposure involved 60% alumina, 25% silica, and 12% iron and magnesium oxides.
- Dyspnea, cough, substernal pain, weakness, and fatigue characterized Shaver's disease, and emphysema-like changes were also noted. Exact role of alumina vs. silica in this disease is difficult to interpret.
- Other reports of aluminum dust/fume/powder exposure are associated with forms of pulmonary fibrosis.
- There is no well-established method for the biological monitoring of aluminum.

Antimony

- Antimony is used as an alloy with lead to make the lead harder and stiffer (as in lead-acid storage batteries). Practically all industrial antimony contains some arsenic.
- Antimony exposure can result in irritation of the skin and mucous membranes. It causes a metallic taste in the mouth (similar to lead), vomiting, colic and diarrhea. Chronic exposure results in loss of appetite, indigestion, and diarrhea. Inhaling oil causes bioaccumulation in fatty tissues.

- Many toxic symptoms are similar to arsenic and lead, but if skin lesions are present, it is generally a sign of antimony exposure.
- When antimony is treated with acid, an extremely toxic, hemolytic gas called stibine is produced.
- Antimony trioxide is considered a probable carcinogen (A2) for lung cancer.
- British anti-lewisite (BAL) is the treatment of choice for antimony exposure. BAL chelating agents bind to metal to help excrete it.
- There is no well-established method of biological monitoring for antimony.

Arsenic

- Chief industrial uses of arsenic include the manufacture of insecticides, weed killers, fungicides, anti-fouling paints, wood preservatives, and semiconductors. Arsenic was once prescribed for treatment of syphilis and for treating parasites.
- Arsenic is a colorless, very poisonous gas with an unpleasant, garlic-like odor. It may be formed accidentally by reaction of arsenic impurities in commercial acids stored in metal tanks.
- In the body, arsenic combines with hemoglobin to form a potent hemolytic agent.
- Arsenic typically acts as a hepatotoxin, causing hemoglobinuria, jaundice, hemolytic anemia, reversible hemolysis, and persistent renal damage.
- Exposures often result in irritation of the skin and conjunctiva of eyes and mucous membrane of the nose. Perforation of the nasal septum is often seen.
- The Ar^{+3} ion is 20 to 60 times more toxic than the Ar^{+5} ion.
- All arsenicals act by inhibiting sulfhydryl enzyme systems required for cell metabolism.

- ❑ Blackfoot disease was observed in the southwest coast of Taiwan as a result from high arsenic concentrations in artesian well water. It affected young adults in their 20s and 30s, but not older individuals. At only 2-3 ppm in the water supply, it produced arteriosclerotic disease in the lower extremities (dry gangrene) of those exposed. This contaminated water may have also contained ergot alkaloids that may have had similar digital gangrene effects.
- ❑ Arsenic is considered a human carcinogen, affecting the skin and upper respiratory system.
- ❑ Arsenic may result in neurological changes (peripheral neuropathy), including numbness, loss of reflex, and muscle weakness.
- ❑ The specific antidote for arsenic and arsine poisoning is dimercaprol, or BAL.
- ❑ Biological monitoring detects recent exposure only. It is not accurate for past exposure. Inorganic arsenic metabolites may be found in urine at the end of the workweek, where BEI is less than 50µg/g creatinine. Some shellfish contain arsenic.

Beryllium

- ❑ The most important occupational beryllium compound is beryllium oxide. It is used as a base material in ceramics and special types of glass, fluorescent tubes, nuclear reactors, and was also used in neon lamp manufacturing in the 1940s and 1950s. When bombarded with alpha radiation, beryllium emits a large number of neutrons. Also is used as a hardening agent in alloys, particularly copper.
- ❑ Beryllium's main target organ of toxicity is the lung. Short-term exposure to high concentrations of more soluble compounds of beryllium can induce chemical pneumonitis, which is often severe and even fatal.
- ❑ Animal studies have defined five disease states associated with beryllium exposure:
  - ❑ Acute chemical pneumonitis.
  - ❑ Chronic pneumonitis with granulomata (called Berylliosis).
  - ❑ Pulmonary tumors, both benign and malignant.
  - ❑ Bone sarcoma.
  - ❑ Rickets and osteosclerosis.
  - ❑ White bands across fingers.
- ❑ Chronic beryllium disease, an immunologic disease, is characterized by granulomas, fibrosis, emphysema, reductions in vital capacity of lungs, and total lung capacity.

- Granulomatous lung disease is debilitating and sometimes fatal.

- Beryllium is a carcinogen (lung cancer, A2).

- Beryllium can be found in urine. Biological monitoring is effective long after exposure has stopped.

Cadmium

- Cadmium is used for plating easily corroded metals such as iron and steel, is found in lead and zinc smelting, silver soldering, paint pigment, and also as a catalyst in plastics and nickel-cadmium batteries.

- Cadmium is predominantly stored in soft tissues (e.g., the liver and kidneys). It has a long biologic half-life (30 years in humans).

- Toxic effects of cadmium include lung damage (emphysema), lung cancer, and kidney damage (marked by Beta-2-microglobulin excretion); it also affects bones, and is suspected in causing metal fume fever, pulmonary edema, and pneumonitis.

- Itai Itai disease (Japan) occurred in postmenopausal women living in cadmium contaminated areas. It was characterized by bone disorders (such as osteomalacia, osteoporosis, increased bone fractures). It was most likely due to the effect of cadmium on calcium metabolism and urinary loss of calcium and phosphates. This resulted in pathologic fractures or bones.

- This metal interferes with several important nervous system functions. It has been linked to cognitive impairment in children and adults, and is an A2 carcinogen, causing lung cancer.

- Biological monitoring can detect recent cadmium exposure in blood, and metallothionien in urine, but both zinc and copper can interfere. ACGIH recommends cadmium urine tests with a BEI less than 5μg/g creatinine or 5μg/L of cadmium in blood.

Chromium

- Chromium is an essential trace element in human nutrition (in its trivalent form), and in the metabolism of glucose.

- Chromium is used in chrome plating. It is alloyed with iron to improve its resistance to corrosion. Genuine stainless steel always contains nickel and chromium (hexavalent chromium is generated in the welding of stainless steel).

- The most important valences of chromium are +3 and +6, although others exist. Chromium +6 (called chromate) is the most toxic (and also the more irritating) and is an A1 carcinogen.
- Industrial exposure causes dermatitis, skin ulcers, and perforation of the nasal septum. Cancer of the lungs, nasal cavity, and sinuses also occurs.
- Chromium is a very common skin sensitizer (especially potassium dichromate, second only to nickel).
- Massive absorption of chromate will occasionally result in cases of acute tubular necrosis (ATN) of the kidneys.
- Excess chromium is excreted within 24 hours. The biological exposure index measures a workshift of 10μg/g creatinine or less, or an end of shift of less than 30μg/g creatinine.

Lead

- The greatest single use of lead today is in plates for storage batteries for cars. Tetraethyl or tetramethyl lead is added to gasoline to improve engine efficiency and reduce gas consumption in automobiles. Due to the effects of lead on health and environment, however, its use in gasoline is being phased out.
- Lead is an infamous cause of childhood poisoning in the following ways:
  - Pica, the habit of eating paint chips by children.
  - The strong effects it has on CNS development in children.
  - May cause mental retardation, failure of speech, aggressive behavior, seizure disorder, and decreased ability in school.
- In adults, toxicity is mainly in the peripheral nervous system. Inhaled lead is fairly well absorbed (35-50%), while ingested lead is not (about 10%). Skin absorption is not a concern, unless tetraethyl lead is involved.
- Occupationally affected workers indicate neurasthenic symptoms as common findings (“wrist-drop” or “foot-drop”). CNS symptoms and muscle and joint pain are also noted.
- Effects on the hematopoietic system of workers are prominent. Symptoms include anemia (caused by impairment of heme synthesis).
- The kidney is the major route of excretion for lead. The kidney is also a site of toxic action of lead (such as lesions of the proximal tubule). As renal function is impaired (either by lead or due to other renal damage), the half-life of lead in the body increases dramatically.

- There is little evidence to show that lead is a carcinogen or co-carcinogen in humans. It has been shown to have immunosuppressive effects in humans.

- Chelation therapy involves the use of calcium disodium EDTA. However, this chelator also has toxic effects on the kidney, so prophylactic use is now prohibited.

- Biological monitoring can be accomplished by testing blood for recent exposure. The BEI should be less than 30μg/100 ml.

Manganese

- Manganese is used in the manufacture of steel and dry cells for batteries. Manganese as an essential element is involved in many enzymatic processes in the body and is involved in metabolism of carbohydrates in general.

- Chronic manganese intoxication manifests neurologic signs resembling Parkinson's disease. The initial phase is somewhat vague. Symptoms include fatigue, irritability, headaches, and asthenia. Symptoms are reversible if exposure is eliminated.

- Chronic exposure to manganese leads to objective signs that may be irreversible. The most characteristic sign of advanced intoxication is a complete absence of facial expressions.

- L-Dopa, used in treating Parkinson's disease, can have beneficial impact on manganese toxicity, but manganese toxicity and Parkinson's are not the same clinically.

- There is no well-established biological monitoring method for checking manganese exposure.

Mercury

- Mercury is often called "quicksilver." Mercury and mercury compounds were used from the 15th to the mid-20th centuries for treatment of syphilis. It is currently used in electrical switches, thermometers, dental fillings, and mercury vapor lamps.

- Mercury released into the environmental may be converted by anaerobic bacteria into methylmercury, then carried through the food chain, particularly to humans by fish, such as in the Minimata Bay of Japan. Nearly all human exposure to methylmercury occurs through eating fish.

- Mercury breath concentrations of people with dental amalgam containing mercury averages 8.2 ng/1, while the mean concentration of those without mercury containing amalgam equals 0.06 ng/1.

- Classic symptoms of exposure to elemental mercury and methyl mercury involve the central nervous system. Classic symptoms of exposure to inorganic salts of mercury (e.g., mercuric chloride) involve kidney function.

- Chronic occupational exposure to mercury vapors can result in tremors, and erethism (insomnia, emotional instability, short-term memory defects, and decreased nerve conduction). This is termed "Mad Hatter's Syndrome."

- Acute effects of inorganic mercury are characterized by gastrointestinal disturbances. Renal failure may also result. Chronic effects include neuropsychiatric consequences, and renal dysfunction, or proteinuria, due to necrosis of the proximal tubule of the nephron.

- Mercuric chloride renal toxicity is distinguished by an initial reduction in renal blood flow.

- The Minimata Bay episode showed teratogenic effects of methylmercury due to the number of children born with cerebral palsy, even though the mothers had minimal symptoms of mercury poisoning during their pregnancies.

- Biological monitoring should be conducted as follows: total inorganic mercury in urine at pre-shift less than 35µg/g creatinine or total inorganic mercury in blood at end of work week less than 15 µg/L.

Nickel

- Nickel is used in the production of nickel metal, nickel containing alloys, and plating. Stainless steel may contain up to 35% nickel.

- Nickel dermatitis (nickel itch) is a common affliction. It consists of both simple dermatitis and chronic eczema, or a neurodermatitis. Dermal exposures may result in a sensitization reaction.

- Nickel is a class A1 carcinogen, causing lung cancer and possibly nasal cancer.

- Blood monitoring within 24 hours of exposure may indicate nickel exposure, yet there is no recommended level of exposure BEI.

Nitroalkanes

Uses and Hazards of Nitroalkanes

- Derivatives of alkanes where one or more hydrogen atoms are replaced by the electronegative nitro group (NO_2). These compounds are acidic substances, with strong solvent power.
- Used in inks, paints, varnishes, and adhesives. Also used as fuel additives, rocket propellants, and explosives.
- Hazards relate to the volatile nature of the compounds, with moderate irritation of mucous membranes, headache, dizziness, and nausea.
- Nitroalkanes are absorbed through the gastro-intestinal tract and lungs, but not through the skin.

Nitropropane

- 2-Nitropropane (2-NP) is one of the more toxic nitroalkanes.
- Some fatalities are attributed to overexposure.
- Fatalities all show destruction of hepatocytes, with liver failure being the primary cause of death.
- Also considered an A2 carcinogen by ACGIH (causes liver cancer in laboratory animals).

Organic Phosphates

Uses and Hazards of Organic Phosphates

- Widely used as insecticides, gasoline additives, hydraulic fluids, cotton defoliants, and fire retardants. Insecticide use of these chemicals is based on their biocide activity, and short life as residues.
- Primary form of action in the body is inhibition of acetylcholinesterase (similar to carbamates). Common examples include diazinon, malathion, and parathion.
- Signs of intoxication reflect stimulation of autonomic and central nervous system (due to accumulation of acetylcholine).
- Early signs of exposure in humans include intestinal cramps, tightness in chest, blurred vision, headache, diarrhea, decreased blood pressure, and salivation. Early symptoms can be counteracted by atropine.

- Organic phosphates as a class generally are not mutagenic (exceptions are dematon, dichlorvos and thio TEPA). Certain organophosphate insecticides are associated with a delayed paralysis. Originally this effect was thought to be due to de-myelination, but it has subsequently been shown to be the result of degeneration of axons in the spinal cord and peripheral nerves.

- The EPA requires greater emphasis on neurotoxic studies of these substances. Effects are apparently not related to cholinesterase inhibition. The effect is always delayed, never acute, and it is structure specific (not all organophosphates show this effect). There is no known effective treatment. Some mild cases slowly return to normal, whereas the severe cases are usually permanent.

- Insecticides are rapidly absorbed through the skin. Skin contact is frequently unnoticed because materials generally do not cause dermal irritation. Inhalation exposure can also occur as a result of inhalation of particulate matter rather than vapor.

Examples of Organic Phosphates

- Diazinon:
  - Oral LD50 in rats = 100–150 mg/kg (very toxic).
  - TLV of 0.1 mg/m^3.

- Malathion:
  - Generally considered the least toxic of the class.
  - Has an oral LD50 in rats of approximately 1400 mg/kg (moderately toxic).
  - TLV of 10 mg/m^3.

- Parathion:
  - Widely used, despite relatively high toxicity.
  - Oral LD50 in rats is between 3.5–12.5 mg/kg (extremely toxic)
  - TLV of 0.1 mg/m^3.

Phenols and Phenolic Compounds

Phenols

- Phenol is sometimes referred to as hydroxybenzene and carbolic acid (C_6H_5OH).

- Used in production of :
  - Aromatic compounds.
  - Explosives.
  - Fertilizers.

- ❑ Paints.
- ❑ Wood preservatives.
- ❑ General disinfectants.

❑ Toxic effects include:
- ❑ Coma.
- ❑ Loss of vasoconstriction.
- ❑ Hypothermia.
- ❑ Cardiac depression.
- ❑ Respiratory arrest.
- ❑ Renal complications.

❑ Corrosive properties can cause:
- ❑ Skin irritation.
- ❑ Mucous membrane and esophageal swelling.
- ❑ Hyperemia in lungs.
- ❑ Bronchitis.
- ❑ Enlargement of hepatic cells.
- ❑ Nephritis.
- ❑ In acute intoxication, death due to respiratory failure.

❑ TLV includes skin notation and has a BEI for creatinine in urine (relatively non-specific, it is seen with other chemical exposures as well).

❑ Once used in local application of dressings or compresses saturated with a dilute (1–5%) solution. No longer used for this purpose due to gangrene, with resultant amputation of fingers and toes.

❑ For cresol and hydroquinone, the toxic effects are similar to phenol. Hydroquinone is actually more toxic than phenol.

Creosote

❑ Creosote can contain cresol, phenol, pyridine, and other aromatics. Creosote is used as an antiseptic, disinfectant, germicide, and wood preservative.

❑ Creosote can induce skin burns, particularly in fair skinned workers. May also cause skin cancer.

❑ Protective garments, petroleum jelly, and other barriers can minimize skin impacts.

Pentachlorophenol

❑ Pentachlorophenol (PCP)—C_6Cl_5OH—can cause dermatitis and systemic intoxication. It was formerly used as a wood preservative.

- Common symptoms are fever, headache, malaise, abdominal pain, and feeling of thirst. Extreme cases include coma, convulsions, vascular collapse, and aplastic anemia with excessive exposure.

Silicon and Silicates

Silicon and Silicates

- Ninety percent of the earth's crust is made up of silica and naturally occurring silicate minerals.
- Silicon compounds are generally poorly absorbed. When absorbed, they are not reactive, and are not toxic in the classic sense of the word.
- Several naturally occurring silicon-containing compounds (e.g., crystalline silica and certain fibrous silicates like asbestos) have resulted in extensive heath effects from benign and malignant respiratory diseases. The physical characteristics of silicon material apparently determine this biologic activity.

Silicon

- Silicon does not occur naturally. It is prepared from silicon compounds (quartz sand or gravel). It is used in metallurgy and also in semiconductor manufacturing.
- Silicon undergoes spontaneous oxidation in air to form a thin layer of silicon dioxide (silica) on a surface.
- Silicon dust is essentially inert. Air exposure limits are set at nuisance dust levels. Dust particles have a surface layer of silicon dioxide.
- At temperatures above 1200°C, the possibility exists that silicon may be converted to more hazardous crystalline silica.

Silicon Dioxide

- Silicon dioxide (silica) comprises 60 percent of the earth's crust as both crystalline silica and amorphous silica. Crystalline silica's most common form is quartz. Other forms include cristobalite, tridymite, and tripoli.
- Silicon dioxide can cause silica-induced pneumoconiosis (silicosis), which has been known by a number of names, including grinders' asthma and miners' asthma.
- Silicon dioxide may also cause nodular pulmonary fibrosis, characterized by accumulation of activated inflammatory cells, granuloma formation, and proliferation of connective tissue. This is typically not diagnosed until 20 years after initial exposure. Symptoms include cough and shortness of breath. Once progressive, it is frequently fatal.

- Fibrosis also increases susceptibility to lung infections. There is an increased risk of tuberculosis (TB) among silicotics.
- Cristobalite has greater fibrogenic potency than quartz and has a lower TLV.
- Crystalline silica is cytotoxic to macrophages and interferes with the normal clearance mechanism of the lungs.
- Evidence of carcinogenic potential of silica in humans is equivocal and difficult to interpret.
- Amorphous silica can produce some pulmonary effects, but at much higher concentrations than crystalline silica. Lesions are rapidly reversible when exposure ceases.

Asbestos

- Asbestos is a naturally occurring material. It is used for thermal, friction, and fire-resistant materials.
- Adverse health effects of asbestos were first noted in the early decades of 20th century.
- Nonmalignant respiratory diseases attributable to asbestos exposure are chronic pulmonary fibrosis (asbestosis) and three pleural diseases (fibrotic pleural plaques, pleuritis, and diffuse pleural thickening). Symptoms include shortness of breath and persistent cough. Exposure is also associated with lung cancer and mesothelioma.
- Epidemiologic evidence clearly shows risks of asbestosis, and lung cancers are significantly greater for smokers.

Talc

- Talc is a naturally occurring silicate. Talc may contain crystalline silica (quartz) and fibrous forms of asbestos.
- Silicosis and asbestosis are frequently seen in talc miners.

Man-made Mineral Fibers

- "Man-made Mineral Fibers" (MMMF) include glass fibers, glass wool, and ceramic fibers. Man-made mineral fibers are used for acoustic and thermal insulation, textiles, and fabrics.
- Fibers larger than five micrometers cause mechanical irritation of the skin as a result of direct contact.
- Studies in animals suggest that MMMF are far less fibrogenic than asbestos fibers. Fibrosis is reversible after exposure ceases.

- Mesothelioma has been noted in lab animals with intrapleural or intraperitoneal administration of MMMF (other than glass filaments).

- Development of mesothelioma appears to be related to the number of fibers less than 0.25 micrometers in diameter and more than eight micrometers in length.

- International Agency for Research on Cancer (IARC) concluded that "sufficient evidence" exists that glass, wool, and ceramic fibers causes cancer in animals and "limited evidence" that rock wool causes cancer. It also concluded that there was "inadequate evidence" that slag wool and glass filaments cause cancer in animals.

- Human epidemiologic data has not been very convincing of the carcinogenic effects in workers employed in manufacturing of glass fibers.

- Neither OSHA nor ACGIH currently consider MMMF to be human carcinogens.

Sulfur Oxides

Sulfur Dioxide

- Sulfur dioxide (SO_2) is formed whenever sulfur is burned in air. It is a widely encountered irritant gas.

- Sulfur dioxide has been associated with acid rain (along with nitrogen oxide).

- Sulfur dioxide is water-soluble and generally does not penetrate to the lower respiratory tract (except during exercise).

- Major health effects are on the upper respiratory tract. Produces sulfurous acid a direct irritant on contact with moist mucous membranes and inhibits mucociliary transport. Bronchoconstriction has been seen in sensitive individuals.

- Sulfur dioxide may act as a cancer promoter (based on high mortality of arsenic smelter workers exposed to SO_2).

Hydrogen Sulfide

- Hydrogen sulfide (H_2S) is found in numerous industries, and is found naturally wherever proteins decompose (e.g., sewer gases, rotten egg odor).

- The toxicity of hydrogen sulfide is often casually disregarded. Hydrogen sulfide's toxic properties are similar to hydrogen cyanide.

- The odor threshold is about 1/400th of the lowest amount that can cause injury. Caution is warranted as olfactory fatigue can cause injury. Odor threshold is about 25 ppm. Levels of 3-5 ppm are also offensive, and at 100 ppm, there is no odor due to loss of olfactory sensation. Levels of 1,000 ppm and higher can be lethal.

- Acute health effects of hydrogen sulfide include mucous membrane irritation, salivation, nausea, headache, and respiratory paralysis, leading to death. It is the leading cause of sudden death in the workplace.

Carbon Disulfide

- Carbon disulfide (CS_2) is primarily used as a solvent. It is a highly flammable liquid, has a foul, slightly ethereal odor, and is very fat-soluble.

- The lungs and skin absorb carbon disulfide. It affects the CNS, cardiovascular system, eyes, kidneys, liver, and skin. Neuropsychiatric signs include mania, hallucinations, memory loss, increased suicide rates, and tremors. Death may result from respiratory failure.

- Chronic exposure may result in degenerative changes in blood and blood-forming organs.

- Epidemiologic evidence from female workers in China suggests teratogenic effects, including congenital heart defects.

- Carbon disulfide may act as a cancer promoter (based on high mortality of arsenic smelter workers exposed to SO_2).

8 Biological Hazards

In the workplace, exposure to biological hazards results in a considerable amount of occupational related disease. Biological agents capable of causing work-related illnesses include infectious microorganisms, biological allergens, and toxins. Industrial hygienists must understand the potential for hazardous agents to cause disease, their pathogenicity, routes of entry, and effective control measures.

Occupations at Risk

THOSE AT RISK

- Workers most at risk to biohazards include agricultural, medical, and laboratory workers.

Biotechnology Workers

- Researchers and production workers of pharmaceutical and biotechnology facilities are at risk for occupational exposure to genetically engineered bacteria, molds, and yeasts.
- Laboratory workers have developed Q fever from infected sheep by inhaling the rickettsial agent, *coxiella burnetii.*
- The hantavirus has infected laboratory workers who have inhaled aerosols produced by infected laboratory animals.

Hospital Workers

- Health care facilities, hospitals, dental offices, blood banks, and other clinics may expose their employees to infectious agents, cytotoxic drugs, anesthetic gases, and other hazards.
- The illnesses acquired may be transmitted from patients to workers or through worker exposures.

Animal Workers

- Veterinarians, zoo workers, taxidermists, and workers in animal product-processing facilities are at risk for occupational exposure to animal-related biological hazards.
- Workers associated with birds such as parrots, pigeons, or parakeets in pet shops, zoos, on construction sites, or in public buildings near perching sites may be exposed to *chlamydia psittaci.*
- Proteins in shed animal dander, hair, animal urine, serum, saliva, or tissues may cause allergic reactions.

Agricultural Workers

- Agricultural workers and those who process agricultural products are exposed to numerous biological agents, such as infectious microorganisms, their spores, and toxins.

- Agricultural workers often suffer allergic or respiratory symptoms in response to mycotoxins associated with fungi such as *penicillium, aspergillus, fusarium,* and endotoxins stored in grains.

- Workers who handle animal products may acquire bacterial skin diseases, such as anthrax, from working with contaminated hides.

- Poultry processing workers' and farmers' exposure to infected turkeys, geese, and ducks, or the aerosolized feces from these birds, may cause psittacosis, a bacterial infection caused by *chlamydia psittaci.*

- Animal-related diseases can be contracted through animal contact, but are more often acquired in the work environment. Risk varies with the type and species of animal.

Other Hazardous Occupations

- Sewage and compost workers may be exposed to enteric bacteria, hepatitis A, and parasitic protozoa such as *giardia.*

- Miners may be exposed to mycobacterium, dermatophytic fungi, and fungi causing deep mycosis.

- Wood-processing workers may be exposed to endotoxins and allergenic fungi, growing on timber.

- Workers in textile manufacturing who process plant fiber may be exposed to gram-negative bacteria that produce endotoxins.

Factors Affecting Infection and Exposure

TYPES OF BIOLOGICAL AGENTS

- There are more than 200 biological agents known to produce infections and allergenic, toxic, or carcinogenic reactions in workers.

- These include:
  - Microorganisms, such as viruses, bacteria, and fungi capable of producing infection, exposure, and allergies.
  - Allergens and toxins generated by plants, that can produce dermatitis, rhinitis, or asthma.
  - Protein allergens from vertebrate animals, such as dander, hair, and saliva.
  - Arthropod-induced illness, such as allergic reaction or inflammation caused by a bite or sting from an insect.
  - Parasite-induced illness such as schistosomiasis caused by the protozoa *Schistoma* or roundworm infestations caused by *Ascaris.*

INFECTION AND EXPOSURE FACTORS

- ❑ Merely working with a disease causing agent or substance does not necessarily produce disease.
- ❑ Many factors contribute to exposure and infection. These include:
  - ❑ Modes of transmission.
  - ❑ Route of entry.
  - ❑ Infectious dose.
  - ❑ Viability and virulence of the agent.
  - ❑ Host susceptibility.

Modes of Transmission

- ❑ There are various modes of transmission for infectious microorganisms and other biological materials. These include contact transmission, vector-borne transmission, and airborne transmission.

Direct and Indirect Contact

- ❑ Direct contact of an infected person may occur in a hospital or medical environment where patients are treated.
- ❑ Zoonotic transmission through bites, stings, and scratches may occur when animals are treated or used in research.
- ❑ Infectious material spilled or splashed directly onto mucous membranes, irritated skin, or open cuts is another direct contact mode of transmission.
- ❑ An indirect contact method of transmission may occur when laboratory equipment or counters are contaminated and the infectious agent is transmitted to a host.

Vector-borne Transmission

- ❑ Diseases may develop when a living vector (e.g., mosquito, flea, tick, etc.) biologically or mechanically transmits a causative agent through a bite directly on the skin.
- ❑ Examples of vector-borne transmission include:
  - ❑ Bubonic plague transmitted via fleas.
  - ❑ Rocky Mountain spotted fever transmitted through ticks and mosquitoes.
  - ❑ Lyme disease transmitted through the deer tick.

Airborne Transmission

- ❑ Infectious diseases may be transmitted through the inhalation of infectious airborne particles. *Mycobacterium tuberculosis* may be transmitted through this means.
- ❑ Workers may also be exposed by contact following the deposition of droplets, through splashing or spilling, onto surfaces, equipment, and personnel.

Routes of Entry

- Occupationally-acquired infections may enter the body through inhalation, ingestion, penetration through irritated skin, and contact with mucous membranes of the nose, eyes, and mouth.
- Ingestion may occur through mouth pipette activities and hand-to-mouth contamination, such as eating, drinking, or smoking in the work area.
- Contaminated objects (such as hypodermic needles, syringes, broken glassware, and scalpels) may inadvertently puncture or the scratch the skin and transmit disease.
- Hand-to-mouth (with or without gloves) or contact with other mucous membranes of the eyes and nose may cause infectious disease. There have been reported cases of HIV infections related to splashes of the eye with infectious agents.

Infectious Dose

- Infectious dose is the number of microorganisms required to initiate an infection.
- The National Institutes of Health (NIH) accumulate and report data related to infectious dose (ID).
- ID^{50} is the number of organisms required to infect 50 percent of a test population.

Viability and Virulence

- A microorganism must be viable (capable of replicating) and the opportunity for infection must exist for infection to occur.
- Temperature, humidity, and the presence or absence of growth factors affect viability.
- *Mycobacterium tuberculosis* is unaffected by drying and remains viable on work surfaces, while most other infectious agents cannot survive these conditions.
- Virulence varies greatly among types and strains of microorganisms. Some microorganisms are highly pathogenic, even in healthy populations. Other microorganisms are opportunistic and are able to infect only hosts with lowered immunities.

Host Susceptibility

- Several medical conditions can increase the risk of infection. These include allergies, pregnancy, skin disorders, antibiotic therapy, and drugs that may alter host defenses.

- Greater use of safe work practices and personal protective equipment may be necessary to protect workers at higher risk.

Hazard Classification

HAZARD CATEGORIES

- Etiologic agents and other infectious materials are considered hazardous materials under the US Department of Transportation (US DOT) 49 CFR Parts 171-180. Etiologic agents and other infectious materials are not included as hazardous agents under the OSHA Hazard Communication Standard.
- The Center for Disease Control (CDC), the National Institutes of Health (NIH), and the National Research Council (NRC) recommend workplace guidelines for containment of biohazardous agents.
- The CDC and NIH have designated four biosafety levels (BSLs) for working with infectious agents or activities using experimentally or naturally infected vertebrates.
- Each biosafety level includes a combination of safe work practices and techniques, safety equipment, and facility design. The combination must be specific for the operations conducted, the agent's route of transmission, and the laboratory activity performed.

Biosafety Level 1 for Infectious Agents

- Biosafety level 1 (BSL-1) is used for work involving no known or minimal potential hazard.
- The agents are not known to cause disease in healthy adults.
- Safe work practices include standard microbiological practices.
- There is no specific safety equipment necessary.
- Work is usually performed on an open bench.

Biosafety Level 2 for Infectious Agents

- Biosafety level 2 (BSL-2) is used for work with moderate-risk indigenous agents associated with human disease.
- These agents may present a hazard from auto-inoculation, ingestion, and mucous membrane exposure.
- BSL-1 practices must be implemented, in addition to the use of biohazard-warning signs, "sharps" precautions, and a biosafety manual defining any needed waste decontamination or medical surveillance policies.

- Safe work practices include primary barriers, including biosafety cabinets (class I or II) or other physical containment devices used for all manipulations of agents that cause splashes or aerosols of infectious materials. PPE, laboratory coats, and face protection are also required as needed.

- Work is usually performed on an open-bench with a sink required and an autoclave available.

- Examples of BSL-2 agents include Hepatitis B virus, *salmonella* spp., and *toxoplasma* spp.

Biosafety Level 3 for Infectious Agents

- Biosafety level 3 (BSL-3) is used for work with indigenous or exotic agents where the potential of infection by aerosols is real and the disease may have serious or lethal consequences. Indigenous and exotic agents vary by country.

- BSL-2 practices must be implemented, as well as controlled access, decontamination of all waste, decontamination of lab clothing before laundering, and a baseline serum.

- Safe work practices include a class I or II biosafety cabinet or other physical containment devices used for all manipulations of agents, PPE, protective lab clothing, gloves, and respiratory protection as needed.

- Facilities must include all those for BSL-2, as well as physical separation from access corridors, self-closing, double-door access, and exhausted air that is not recirculated and negatively airflowed into the laboratory.

- Examples of agents requiring BSL-3 include *mycobacterium tuberculosis*, and *coxiella burnetii*.

Biosafety Level 4 for Infectious Agents

- Biosafety level 4 (BSL-4) is used for work with dangerous and exotic agents that pose a high individual risk of life-threatening disease, aerosol-transmitted lab infections, or related agents with unknown risk of transmission.

- Safe work practices include BSL-3 requirements as well as changing clothes before entering, showering before exiting, and decontaminating all material on exiting the facility.

- Safety equipment includes all procedures conducted in class III biosafety cabinets or combination class I or II biosafety cabinets with full-body air-supplied, positive-pressure personnel suits.

- ❑ Facilities require BSL-3 plus a separate building or isolated area, dedicated supply and exhaust air, vacuum, and decontamination systems.
- ❑ Examples include the Ebola, Lassa, and Marburg viruses.

Hazard Control

PREVENTION AND CONTROL

- ❑ Controls to prevent occupational exposure to infectious agents are achieved through source control, minimization of accidental releases, and worker protection.

Containment

- ❑ Containment is a significant control method for protecting workers, the work environment, and the community from exposure.
- ❑ Facility design, safety equipment, and safe work practices are important factors of containment.
- ❑ Containment is achieved through a combination of primary barriers such as class I, II, or III biosafety cabinets and the use of personal protective equipment.

Facility Design

- ❑ Facility design is an important factor in containment. A facility may be designed with the laboratory as a separate building or in an isolated zone with its own dedicated supply and exhaust heating, ventilating, and air-conditioning system.

Safety Equipment

- ❑ Biological safety cabinets are used frequently as laboratory containment equipment, which provides a primary barrier to prevent escape of infectious aerosols into the work environment.
- ❑ The cabinet functions similar to a chemical lab hood, except for additional high-efficiency-particulate-air (HEPA) filters that are present for exhaust air.
- ❑ The cabinets are rated in effectiveness from I, II, and III. Class I is relatively rare, yet may be used for containment of aerosol-producing equipment such as centrifuges.
- ❑ Class II biosafety cabinets provide protection to workers, experimental materials, and the environment through a front opening.
- ❑ Class III biosafety cabinets are completely enclosed, ventilated, provide negative-pressure, and are of gas-tight construction. They are used for work requiring the highest level of containment.

Work Practices

- The most critical component in preventing occupational exposure is safe work practices. Seven basic safe work practices are often required in laboratory settings.

- These minimum practices are:
  - Mouth pipetting is prohibited.
  - Manipulate infectious fluids carefully to avoid spills and the production of aerosols and droplets.
  - Restrict the use of needles and syringes to procedures for which there are no alternatives; use needles, syringes, and other sharps carefully to avoid self-inoculation; and dispose of sharps in leak and puncture resistant containers.
  - Use protective laboratory coats and gloves.
  - Wash hands after all laboratory activities, after removing gloves, and immediately following contact with infectious materials.
  - Decontaminate work surfaces before and after use and immediately after spills.
  - Do not eat, drink, store food, apply cosmetics, or smoke in the laboratory.

Decontamination

- Decontamination is the physical or chemical method used to clean or otherwise render materials safe for further handling by reducing the number of organisms present.

- Sterilization completely eliminates all forms of microbial life. Steam at 250° Fahrenheit under pressure in an autoclave is the most widely used and convenient method of sterilization.

- Disinfection is the process that kills infectious agents outside the body.

- Chemical disinfectants inactivate microorganisms by one or more of a number of chemical reactions. The level of effectiveness is altered or changed in the concentration of active ingredients, contact duration, temperature, and humidity and the concentration of organic matter and the pH of the material being disinfected.

- Chemical disinfectants include halogens, acids, alkalis, and alcohols, among others. The most common disinfectants are sodium hypochlorite (i.e., household bleach), isopropyl alcohol, and phenolics such as Lysol and amphyl.

Infectious Waste

- Infectious waste is usually treated by autoclaving before disposal into a sanitary landfill, by incineration, or by chemical disinfection.

Biosafety Programs

PROGRAM MANAGEMENT

- Biosafety programs are necessary to effectively assure that workers, their colleagues, and the community are not adversely affected by potentially hazardous microorganisms or their toxins.

- Biosafety program components include:
  - Program support of top management.
  - A biosafety officer or technical specialist.
  - A biosafety committee.
  - A biosafety manual of written policies and procedures.
  - Medical surveillance for relevant employees.
  - Employee training.

Bloodborne Pathogens

BLOODBORNE PATHOGENS

- Bloodborne pathogens are microorganisms that may be present in human blood and body fluids, and are capable of causing disease in humans.

- The development and implementation of measures to promote worker protection from infectious agents was prompted by the hepatitis B virus and the human immunodeficiency virus (HIV), which is known to cause acquired immune deficiency syndrome (AIDS).

- OSHA implemented the regulation of work environments involved with potentially infectious agents with the Bloodborne Pathogen Standard. The standard applies to all occupations in which there is potential for exposure to human blood.

- HIV and hepatitis B and C are most frequently associated with occupational infectious agents, yet exposure to other microorganisms such as malaria and syphilis may also cause bloodborne work-related illness.

Epidemiology

- Hepatitis B virus (HBV) infects liver cells and causes liver damage. Symptoms include flu-like illness, jaundice, extreme fatigue, fever, nausea, and joint pain. HBV presents an occupational risk to persons whose work activities involve handling human blood and body fluids.

- Hepatitis C virus (HCV) is similar to HBV.

- Human immunodeficiency virus (HIV) is transmitted through exposure to human blood and certain body fluids. Infection can result from mucous membrane exposure (eyes, nose, and mouth) to contaminated materials. Infection can result from punctures in the skin from hypodermic needles containing contaminated blood or body fluids.

THE OSHA BLOODBORNE PATHOGEN STANDARD

- The intent of the OSHA Bloodborne Pathogen Standard is to prevent or minimize parenteral nonintact skin and mucous membrane exposure to human blood and body fluids in the work environment.
- Those occupations impacted by this regulation must develop a method to identify employees at risk, and develop an exposure control plan detailing how they will comply with the regulation.
- Key components in prevention of work-related exposure are "universal precautions."
- Universal precautions require that all human blood and certain human body fluids be treated as if known to be infectious for HIV, HBV, and other bloodborne pathogens.

Controls

- The regulation requires engineering controls—specifically, needle disposal containers and equipment for reducing aerosols and splatter.
- The regulation mandates work practice controls with special emphasis on hand washing and personal protective equipment, such as gloves, face protection, and eyewear.
- Immunization is a critical component of bloodborne pathogen hazard control. An effective vaccine is available for HBV. The regulation mandates employers offer the HBV vaccination to employees whose work activities bring them into contact with bloodborne pathogens.
- Sharps management is heavily emphasized and requires a significant number of specialized handling procedures and safe work practices.

Sharps Management

- Heavily emphasized in the OSHA standard, sharps management requires a number of safe work practices and procedures to prevent workplace exposures.
- These practices include:
  - Needle handling procedures, such as not resheathing needles by hand.
  - The use of rigid sharps disposal containers.
  - Labeling and transport requirements.
  - Housekeeping, including disinfection.
  - Disposal of infectious waste.
  - Spill management.
  - Handling of contaminated laundry.
  - Communication of hazards to employees.
  - General workplace practices and procedures.

Tuberculosis

EPIDEMIOLOGY

- The *mycobacterium tuberculosis* causes tuberculosis (TB), a bacterial disease.
- In the work environment, health care workers who provide patient care are at risk of aerosol-borne infectious droplet nuclei from patients. High-risk groups include funeral directors and health care service workers, such as nursing aides, orderlies, and attendants.
- Transmission occurs through exposure to bacilli in airborne droplet nuclei from sputum of persons with infectious tuberculosis.
- Symptoms include fatigue, fever, and weight loss, coughing, and blood-tinged sputum early in the disease. The initial infection usually goes unnoticed. Tuberculin sensitivity appears within a few weeks. Lesions commonly heal, leaving no residual changes except pulmonary calcifications. Progressive pulmonary tuberculosis generally arises from a latent focus remaining from the initial infection.
- With treatment, communicability can be eliminated in several days or a few weeks.

Controls

- Containment for activities with potential TB exposure. The type of activity performed determines biosafety levels. The American Thoracic Society has identified specific activity levels and assigned biosafety level 2 or biosafety level 3 to these activities where TB is a potential exposure hazard.
- PPE (respirators) are required by the Centers for Disease Control (CDC) for health care workers when:
  - Entering the room of a patient in isolation for TB or suspected TB.
  - Present during cough-inducing or aerosol-generating procedures on such patients.
  - In settings where administrative and engineering controls cannot ensure safety.
- Safe Work Practices reduce transmission by following infection control guidelines. The CDC has published specific guidelines for preventing the transmission of TB in health care facilities, extending them to include protection of both patients and personnel.

Legionnaires' Disease

LEGIONELLOSIS

- Legionellosis is a term for any *legionella* infection.

- Legionnaires' disease is an acute bacterial disease with two currently recognized distinct manifestations: Pontiac fever and Legionnaires' pneumonia. Both are characterized initially by anorexia, malaise, myalgia, and headache. Within a day, there is usually a rapidly rising fever associated with chills. A nonproductive cough is common; abdominal pain and diarrhea also occur in many patients.

- People usually recover from legionnaires' disease in a few weeks and suffer no long-term consequences. Antibiotic treatment reduces mortality. Pontiac fever is less serious, with flu-like symptoms. It affects a wider range of people.

Sources and Transmission

- Legionellae are common in nature and found in rivers and lakes.

- Legionellae colonies proliferate in water environments:
  - Where the water is warm (between 77 and 108° Fahrenheit).
  - Where the pH is between 2.5 and 9.5.
  - Where the water is allowed to stagnate, followed by agitation.
  - In the presence of other organisms, sediment, and scale.

- Legionnaires' disease is transmitted via inhalation of airborne bacteria droplets.

- Sources of exposure include sprays from cooling towers or evaporative condensers and humidifiers.

Control

- Good design and system maintenance of ventilation, humidification, and water-cooled heat-transfer equipment is essential to prevent the growth of these bacteria in water systems and equipment and to prevent the spread of airborne bacterial droplets through ventilation systems.

Air Sampling for Biological Agents

SAMPLING EQUIPMENT AND PROCEDURES

- Air samplers are commonly used indoors to collect culturable bioaerosols, including the slit, multiple-hole (sieve) and centrifugal impactors, impingers, cyclones, and cassette and high volume filters.

- Impaction occurring directly onto an agar-based culture medium is the most commonly chosen method for collecting culturable bacteria and fungi.

- Samples that are usually collected include:
  - Outdoors, near supply air intakes.
  - Indoors, in the environments of workers experiencing symptoms or where contamination is suspected.
  - Indoors, in comparison or control locations such as in noncomplaint areas or sites where no contamination was observed.

- Sampling is best conducted at various times of the day and on multiple days. The number of samples necessary depends upon the analytical method.

Interpretation of Results

- When biological materials are found in air samples from test areas but not in those from the control locations, explanations should be sought.

- Higher levels of bacteria indoors than outdoors do not imply a potentially hazardous situation. Human-commensal bacteria such as *Micrococcus* and *Staphylococcus* spp., which people shed from their skin and hair, will be predominantly found indoors.

- Higher levels of environmental bacteria, such as *Bacillus* spp., should prompt a search for an indoor source, such as damp carpets, walls, or papers.

9 Ionizing Radiation

An industrial hygienist must understand the nature of radiation, the detection of radiation, permissible exposure limits, biological effects of radiation, monitoring techniques, procedures, and control measures.

Types, Uses, and Biological Effects of Ionizing Radiation

IONIZING RADIATION

- ❑ Ionizing radiation is any radiation capable of producing ions by interaction with matter.
- ❑ Direct ionizing particles are charged particles (e.g., electrons, protons, alpha particles, etc.) having sufficient kinetic energy to produce ionization by collision.
- ❑ Indirectly ionizing particles are uncharged particles (e.g., neutrons, photons, etc.) that can liberate directly ionizing particles.

Types of Radiation

- ❑ Alpha (α):
  - ❑ Composition: two protons and neutrons (due to nuclear decay).
  - ❑ Source: Ra-226 → Rn-222 → Accelerators.
  - ❑ Particle.
- ❑ Beta (β):
  - ❑ Composition: High energy electron.
  - ❑ Source: Sr-90 → Y-90 → Electron beam machine.
  - ❑ Particle.
- ❑ Gamma (γ):
  - ❑ Composition: Photon (energy generated in nuclear decay).
  - ❑ Source: Tc-99.
  - ❑ Not a particle (electromagnetic wave).
- ❑ X-Ray:
  - ❑ Composition: Photon (generated by electrons leaving an orbit).
  - ❑ Source: Most radioactive materials, X-ray machines, secondary to β.
  - ❑ Not a particle.

Industrial Uses of Ionizing Radiation

- ❑ Uses of ionizing radiation include:
  - ❑ Coating thickness measurement.
  - ❑ Liquid level determination.
  - ❑ Density and mass measurements.

- Static electricity control, such as static masters and static brushes (contains Po-210, an alpha emitter, so highly charged that it attracts dust particles).
- Surface curing.
- Sterilization (e.g., sterilizing hospital equipment).
- Non-destructive testing.
- Gas chromatography (sealed sources in GCs).
- Vacuum measurements.
- Oil well logging.

- Industrial use does not include reactors, weapons, and medical diagnosis, therapy and research.

Biological Effects of Ionizing Radiation

- All radiations cause essentially the same biological effects and act through the same injury mechanism.

- Differences are in the degree and distribution (e.g., fast reproducing cells affect the lining of the stomach and bone marrow).

- Interaction is at the molecular level:
  - Ions created are highly active biochemically.
  - At high levels, cell death results. This is advantageous since DNA is not repaired incorrectly, which would result in birth defects or cancer.
  - The most sensitive bio-molecule is DNA.
  - At exposure levels not causing cell death, DNA damage may result in cancer, birth defects, or genetic effects.

- Repair mechanisms are known to exist and the extent of repair may depend on dose rate and density of irradiation.

- Sensitivity of tissue is related to the rate of cell division and future number of cell divisions.

Internal Radiation Protection

- Controls for inhalation hazards include:
  - Engineering controls (fume hoods, glove boxes, and filtration).
  - Respiratory protection (masks, supplied air, SCBA (good for particulate radiation only)).
  - Monitoring (air sampling and bioassay).

- Ingestion hazard controls include work rules (e.g., controlled areas, clothing, etc.) and contamination control (e.g., monitoring, control points, etc.).

External Radiation Protection

- The three mechanisms for external radiation protection include time, distance, and shielding.

- Time: A shorter time in a radiation field means less dose.

- Distance: From a point source, dose rate is reduced by the square of the distance and expressed by the inverse square law:

$$\frac{I_1(d_1)^2}{(d_2)^2} = I_2 \quad \textbf{or} \quad \frac{I_1}{I_2} = \frac{(d_2)^2}{(d_1)^2} \quad \textbf{or} \quad I_1(d_1)^2 = I_2(d_2)^2$$

where:

I_1 = dose rate or radiation intensity at distance d_1
I_2 = dose rate or radiation intensity at distance d_2

- Shielding: Radiation is reduced exponentially by thickness of shielding material.

Removal of Radioactive Material from the Body

- In the event that radiation has been inhaled or ingested, there are two mechanisms that remove the radiation from the body: normal radioactive decay and biological processes that remove material (e.g., skin growth).

- The rate at which this occurs depends on chemical properties, varies significantly between people, and often follows half-life kinetics.

Effective Half-Life Calculation

- The formula outlined below is used when an industrial hygienist is interested in finding out how much radiation is left in a worker's stomach after a period of time.

- Effective half-life is a combination of radiological and biological half-lives and is expressed as:

$$T_{eff} = \frac{(T_b)(T_r)}{T_b + T_r}$$

where:

T_b = biological half-life
T_r = radiological half-life

- T_{eff} will always be shorter than either T_b or T_r.

- T_b may be modified by diet and physical activity.

Alternate Method for Effective Half Life Calculation

- Another method to estimate effective half-life is:

$$\frac{1}{T_{eff}} = \frac{1}{T_{rad}} + \frac{1}{T_{bio}}$$

where:

T_{eff} = effective half-life
T_{rad} = radiological half-life
T_{bio} = biological half-life

Alpha Radiation

CHARACTERISTICS OF ALPHA RADIATION

- Alpha radiation (α) consists of a stream of alpha particles, identical to the helium nucleus.
- Alpha radiation is emitted spontaneously during the radioactive decay of certain radionuclides, primarily those of high molecular weights (bismuth and higher).
- Alpha radiation's energy is approximately 5MeV (Millielectron volts). This is low energy compared to other ion forms.
- The mass of alpha radiation is 4 Atomic Mass Units (AMUs). It contains two protons and two neutrons ($^{4}_{2}He$) and its charge is positive two (+2).
- The range for alpha radiation is approximately 3-4 cm in air and 5 x 10^{-3} cm in tissue. It does not penetrate matter readily. It is usually stopped by skin. It is more of an internal radiation hazard since it can be inhaled in particulate form.
- Energy is lost through energy excitation and ionization. Each ionization requires about 35 electron volts (eV); each alpha particle reduces about 150,000 ion pairs over its range.

Alpha Radiation Applications

- Alpha radiation is used for:
  - Air ionization—elimination of static electricity (Po-210), clean room applications, and smoke detectors (Am-241).
  - Air density measurement—vacuum gauges (Ra-226, Am-241).
  - Moisture meters, non-destructive testing, and oil well logging.
  - Gamma fluorescence and gamma backscatter gauges.
  - Naturally occurring alpha materials used for physical or chemical properties, including uranium (coloring of ceramic glaze, shielding) and thorium (high temperature materials).

Alpha Emitting Isotopes

- The table below outlines some naturally-occurring isotopes, their radiological half-life, other radiation types emitted, and body burden.

Isotope	T½	Other Radiation	Body Burden (microcuries)
Am-241	400 YEARS	γ, χ	0.05
Po-210	138 DAYS	χ	0.03
Natural U (U-238)	4.5×10^9 YEARS	β, γ, χ	0.005
Natural Th (Th-232)	1.4×10^9 YEARS	β, γ, χ	0.04
Cm	17.6 YEARS	χ	0.1
Ra-226	1602 YEARS	β, γ, χ	0.1
Rn-222	3.8 DAYS	β, γ, χ	NA

ALPHA DETECTION INSTRUMENTATION

- Types of equipment used to evaluate alpha radiation in the workplace include:
  - Geiger-Mueller counter.
  - Gas proportional devices.
  - Scintillators.
  - Solid-state analysis.
- All have high sensitivity. The gas proportional devices can be used to discriminate α from β, and γ.

Gas Proportional

- Gas proportional devices are less affected by background radiation, they are more complex to use, and they contain a large detector.
- These units are portable, of moderate cost, and are good for source monitoring.

Geiger-Mueller

- The Geiger-Mueller (GM) counter contains a pancake shaped probe, and is affected by background radiation. It is simple to use, is low cost, and contains a small detector.
- These units are good for area monitoring only. They can detect very small amounts of radiation.
- GM units are accurate only for the type of radiation for which they are calibrated.

Scintillators

- Zinc sulfide scintillators and scintillation counters (handle probe) are less affected by background radiation, are moderately complex, and contain a small detector.
- These units are good for source measurements, have excellent portability properties, and are of moderate cost.
- These units measure light flashes created by interaction of ionizing radiation and scintillator materials.
- They are excellent for sensitive measurements of alpha, beta, and gamma radiation.

Solid State

- Solid state analysis is less affected by background radiation. It is highly complex, contains a small detector, has poor portability, and is expensive.

EXPOSURE HAZARDS OF ALPHA RADIATION

- The exposure hazards of alpha radiation include both external and internal hazards.
- External exposure factors include:
  - Dead skin layer provides shielding.
  - Does not penetrate deep into the skin.
  - Of concern in certain unusual cases.
- Internal exposure includes:
  - Inhalation: Respirable—particle size determines hazard.
  - Ingestion: Minimal hazard unless absorbed by GI tract. If absorbed, may remain in body a long time ("bone seekers").
  - Wounds: Historically have been involved in some of most serious exposures. May be difficult to assess.

Beta Radiation

CHARACTERISTICS OF BETA RADIATION

- Beta radiation (β) consists of a stream of beta particles, either electrons (negative charge) or positrons (electrons with a positive charge). Positrons are emitted by an atomic nucleus or by a neutron in the process of transformation.
- Beta radiation energy is 1 KeV to 4 MeV (higher than alpha radiation).
- Beta radiation mass is 5.4×10^{-4} AMU. Beta radiation has a negative charge.

- The range of beta radiation approximates 12 feet/MeV in air, 0.5 cm/MeV in tissue, and 70 KeV to penetrate a dead skin layer.

- As beta radiation loses energy, Bremsstrahlung radiation is formed.

Bremsstrahlung Radiation

- Bremsstrahlung radiation (Brem) is the electromagnetic radiation associated with the deceleration of charged particles.

- The term also applies to the radiation associated with the acceleration of charged particles.

- As β particles get close to a nucleus, they decelerate and form x-rays. X-rays are generated by these β sources.

- The percent of electron energy lost to Brem is proportional to the atomic weight of the material.

- Bremsstrahlung radiation is sometimes referred to as "braking radiation."

Beta Radiation Applications and Sources

- Beta radiation is are used for:
  - Thickness measurements for coating operations.
  - Radioluminous signs.
  - Tracers for research.
  - Air ionization (gas chromatograph, nebulizer).

- Naturally occurring beta materials include:
  - Potassium 40.
  - Carbon 14—used in tracer studies.
  - Hydrogen 3—tritium used for static measurement meters and gauges.
  - Strontium 90 and Krypton 85 used for beta thickness gauges.

BETA DETECTION INSTRUMENTATION

- Types of equipment used to evaluate beta radiation in the workplace include:
  - Geiger-Mueller counter.
  - Gas proportional devices.
  - Scintillators.
  - Ion chambers.
  - Dosimeters.

- The requirements for beta detection instruments include the measurement of dose rate (Rem/Hr) as well as counts/minute. Instruments require a thin entrance window in the unit and the ability to discriminate gamma radiation.

- Workers measuring exposure frequently wear dosimeters, including film badges or thermoluminescence devices.

Ion Chamber

- An ion chamber consists of two plates with an electrical potential between them that are placed in a container filled with air. Radiation enters and a current flows in the circuit and becomes conductive.
- These units measure ionization directly and are energy independent.
- These units are good for beta, gamma, and, with a thin window, alpha detection.
- They produce stable responses over a wide voltage range, and have low sensitivity, which is best for high radiation fields.

Film Badge

- A film badge dosimeter is worn on outer clothing as a personal radiation monitor for gamma, x-ray, and high-energy beta radiation.
- A film badge contains photographic film wrapped in a cover.
- It is worn as a badge or as a finger ring.
- Radiation reacts with silver atoms in the photographic film to expose the film.
- The badge is removed and the film is developed.
- The amount of darkening of a film is then compared to a control film that was not exposed during the same period.
- It provides workers with a permanent record of dosage.

Thermoluminescence Detectors

- Another type of dosimeter is the thermoluminescence detector (TLD).
- TLDs are used for gamma, x-ray, and beta exposure monitoring.
- TLDs are worn as a body badge or a finger ring and consist of a small lithium fluoride chip.
- No energy source is required from 20 KeV and up for x-rays and gamma rays.

- The energy absorbed by the chip heats the electrons and displaces them from their ground state. Light is emitted when electrons return to the ground state. The amount of light released is related to the absorbed radiation dose and to the radiation exposure of the worker.

BIOLOGICAL EFFECTS OF BETA RADIATION

- Beta radiation presents two potential exposure methods, external and internal.
  - External beta radiation hazards are primarily skin burns.
  - Internal beta radiation hazards are similar to alpha emitters.

Shielding of Beta Radiation

- Shielding of beta radiation is best accomplished by using low z materials to reduce Bremsstrahlung radiation. A low z material is a material with a low atomic number.

- The thickness is critical to stop maximum energy range, and varies with the type of material used.

- Typical shielding material includes plastics, aluminum, Plexiglas, lead, water, wood, cement, and wax (wax is a good material for beta).

- A shield may require lead or other x-ray shield on the exterior of the β shield to eliminate Brem.

Formula for Calculating Appropriate Shielding for Beta Radiation

- Industrial hygienists use the formula below to determine what a specific shielding thickness should be:

$$I = I_o e^{-\mu x}$$

where:

I = dose rate or radiation intensity shielded (rem/hr)
I_o = original or initial dose rate or radiation intensity unshielded (rem/hr)
x = thickness of shield given for different materials and energies
μ = provided through a table or chart data for the type of material used

Gamma Radiation

CHARACTERISTICS OF GAMMA RADIATION

- The electromagnetic spectrum encompasses gamma (γ) and x-radiation spanning an energy range from approximately 3×10^3 eV to 10^7 eV.

- Gamma radiation is a photon, not a particle. There is no charge, there is no mass, and the range is indefinite.

- Gamma rays are nuclear in origin, with very specific energy. X-rays originate in electron energy transitions without specific energy.

Gamma Radiation and X-Ray Uses

- Gamma radiation is used for:
  - Sterilization of food and medical products.
  - Radiography of welds, castings, and assemblies.
  - Gauging of liquid levels and material density.
  - Oil well logging.
  - Material analysis.

- X-ray machines are used for thickness gauging and radiography, and are secondary to many high voltage devices and electron microscopy.

Gamma and X-Radiation Interactions with Matter

- There are two basic processes by which electromagnetic radiation interacts with matter:
  - Scattering, in which the direction of the photon and its energy are altered.
  - Absorption, in which the photon disappears with transfer of its energy to other radiation.

Absorption

- Absorption of photons occurs primarily by three processes: the photoelectric effect, the Compton effect, and pair production.

- Photoelectric effect predominates for lower energy photons (<100 KeV). Involves complete absorption and interaction between incident photons and the electrons in the shells around the nuclei.

- Compton effect or Compton scatter predominates with greater energies (100-1000 KeV). Involves interactions of photons incident on orbital electrons. The photon gives up part of its energy to the electron, causing it to recoil, and the balance of its energy goes into a scatter photon.

- Pair production occurs in high energies (>1MeV). It involves the interaction of a photon with the electric field surrounding a charged particle. The original photon disappears with the formation of an electron-positron pair.

GAMMA DETECTION INSTRUMENTATION

- Ion chamber:
  - Offers stable response over wide voltage range.
  - Has an energy independent response.
  - Low sensitivity (best for high radiation fields).

- Gas proportional:
  - Can discriminate between α, β, and γ.
  - High sensitivity.
  - Requires stable electronics to maintain voltage.

- Geiger Mueller:
  - Has maximum sensitivity.
  - Has a stable response over a wide voltage range.

BIOLOGICAL EFFECTS OF GAMMA AND X-RAYS

- The principal health concern associated with gamma radiation is external exposure by penetrating radiation and physically strong source housings.

- Sensitive organs include the lens of the eye (causing cataracts), the gonads (genetic effects), and damage to the bone marrow (inducing leukemia).

- Short term effects (i.e., high doses in a short time) include:
  - <50 Rem—Subclinical.
  - 50-500 Rem—Hematopoetic system effects.
  - 500-100 Rem—Gastrointestinal system effects (fatal).
  - >1000 Rem—Central nervous system effects (fatal).

- Long term effects based on high dose data include carcinogenesis and birth defects.

GAMMA AND X-RAY SHIELDING

- Shielding gamma and x-radiation depends on energy level.

- Protection follows an exponential function of shield thickness.

- Shielding may be approximated using tenth or half-value thickness (HVL).

- At low energies, photo-electric absorption can be achieved with just millimeters of lead.

- At high energies, shielding modifies the energy spectrum. Half-value thickness provides estimates and may involve large quantities of lead, water, iron, or concrete. Designing cost effective shielding requires sophisticated calculations.

- Shielding can attenuate gamma radiation. The amount of shielding required depends upon the emission rate of the source and the energy spectrum of the gamma radiation.

Shielding: Half-Value Layer Method

- Using the concept of half-value layers (HVL) in determining the amount of shielding material required is the most common method for calculating the thickness of a specific material for attenuating gamma radiation.

- Use the following formula to calculate shielding based on the half-value layer of the material proposed.

$$X = \frac{\text{Log}\,(ER_o/ER_d) \times HVL}{\text{Log}\,2} \quad \textbf{or} \quad ER_d = ER_o\,(0.5)^{x/HVL}$$

where:

X = thickness of shielding in cm
ER_o = emission rate (observed)
ER_d = emission rate (desired)
HVL = Half-Value Layer

Gamma Dose Rate Calculation

- To calculate dose rate in rads per hour related to energy, activity, and gamma rays/disintegration, use the following formula:

$$D = 6\,CEN \text{ at one foot}$$

where:

C = Curies
E = energy in MeV
N = # of photons/disintegration
D = dose rate in rad/hr

or

$$D = \frac{\Gamma A}{d^2}$$

where:

D = absorbed dose in Rad per hour
Γ = gamma ray constant (R-cm^2/hr-mCi) must be given
A = activity in millicuries
d = distance from a point source

Decay Equations and Units of Activity

BASIC FORMULA FOR DECAY CALCULATIONS

- Radioactive materials decay (emit alpha particles, beta particles, and photon energy), and lose a proportion of their radioactivity with a characteristic half-life.

- To calculate the amount of radioactivity remaining after a given period of time:

  later activity = (earlier activity) e $^{-\lambda}$ (elapsed time)

  $A = A_i e^{-\lambda t}$

where:

$$\lambda = \frac{LN2}{T}$$

and:

λ = lambda decay constant (probability of an atom decaying in a unit time)
t = time
LN2 = 0.693
T = radioactive half-life (time period in which half of a radioactive isotope decays)
A = new or later radioactivity level
A_i = initial radioactivity level

Determining Time to Decay

- Determining the time required for a radioactive material to decay (A_o to A) is expressed as:

$$t = -LN\frac{A}{A_i} \bullet \frac{T}{LN2}$$

where:

λ = lambda decay constant (probability of an atom decaying in a unit time)
t = time
LN2 = 0.693
T = radioactive half-life (time period in which half of a radioactive isotope decays)
A = new or later radioactivity level
A_i = initial radioactivity level

- Rule of Thumb: In seven half lives, reduced to <1%; 10 half lives <0.1%

Rate of Radioactive Decay

- Radioactive material disintegration is directly proportional to the number of nuclei present. The radioactive decay rate is expressed in nuclei disintegrated per unit time. Decay can also be expressed in time as described above.

$$A_i = \frac{0.693}{T} \bullet N_i$$

where:

A_i = initial rate of decay
N_i = initial number of radionuclei
T = half life

Half-life Calculations

- Half-life is defined as the time it takes for a material to lose 50% of its radioactivity.

$$A = A_i(0.5)^{t/T}$$

where:

A = activity at time t
A_i = initial activity
t = time
T = half-life

- Note: The formulas expressed in the last three blocks are similarly derived and solved in a variety of ways.

Units of Activity

- Curie (Ci): The quantity of radioactive material which results in 3.7×10^{10} disintegrations per second (1 G of Ra-226).

milliCurie	10^{-3} Ci
microCurie	10^{-6} Ci
nanoCurie	10^{-9} Ci
picoCurie	10^{-12} Ci

- Becquerel (Bq): 1 disintegration/second, commonly used in Europe.

Decay Problems

1. What will be the activity of 5 Ci of I-131 in 30 days?

$T^{1/2} = 8$ days $\qquad A = 5\,(e^{-0.693/8})\ 30$

$$A = 5(0.74)$$

$$= 370 \text{ mCi}$$

2. How long will it take for 20 Ci of Po-210 to decay to less than 1μCi?

$$T^{1/2} = 138 \text{ days}$$

$$t = -LN \frac{1 \times 10^{-6}}{20} \frac{138}{0.693}$$

$$= 3{,}350 \text{ days or } 9.2 \text{ years}$$

Radiation Dose

RADIATION DOSE UNITS—RAD

- Extensive work and research has been done in trying to relate radiation dose to its resulting damage. Unfortunately, much knowledge has been based on animal studies or human tragedies.

- The rad is one unit of measure associating absorbed dose with resulting damage. The rad is a measure of energy deposited. Rads have different biological interactions with tissue.

- Rad is an synonym for radiation absorbed dose. One rad = 100 ergs/g.

- Quality factors need to be applied to rad measurements in order to get the more useful unit, Roentgen equivalent man or Rem.

RADIATION DOSE UNITS—REM

- Roentgen equivalent man (rem) is a unit of absorbed dose (in rad) times a quality factor (QF, obtained from a table) that is used to express the relative biological effect of the particular radiation, as compared to gamma-radiation.

- The rem is a measure of biological effects. Personnel exposure limits are often expressed in rem. The rem of absorbed radiation causes the equivalent biological damage in tissue regardless of the source (e.g. alpha, beta, or gamma radiation).

- One rem = rad x QF.

- A dose of 500 rem in a short period of time would probably be lethal.

RADIATION DOSE UNITS—QUALITY FACTORS

- The quality factor (QF) reflects the variation of effectiveness of different radiations in causing biological effects.

- ❑ QF is determined by experimentation and varies with the biological end point used.

Quality Factor Table

X-Rays	1.0
Gamma	1.0
Beta (fast)	1.0
Alpha	20.0
Fission fragments	20.0
Neutrons (slow)	10.0
Neutrons (fast)	3.0

Exposure Limits

EXPOSURE LIMITS

- ❑ Occupational exposure limits include the following:
  - ❑ Whole body: 5 rem/year.
  - ❑ Lens of eye: 15 rem/year.
  - ❑ Shallow dose: 50 rem/year.

- ❑ Non-occupational exposure is 0.1 rem/year.

- ❑ Pregnant workers must remain within 0.5 rem/term (entire length of pregnancy). Pregnancy must be officially declared in writing before an employer is required to recognize pregnancy.

- ❑ There is no limit on medical exposures, since a patient is choosing to have x-rays.

10 Non-ionizing Radiation

Non-ionizing radiation is electromagnetic radiation that does not cause ionization. It includes ultraviolet, visible light, laser, infrared, microwave, and radiofrequency radiation. Industrial hygienists must understand the principles of electromagnetic radiation, its uses in the workplace, its hazards and effective control measures.

Non-ionizing Radiation

NON-IONIZING RADIATION AS ENERGY

- Radiation is simply a form of energy.
- Non-ionizing radiation travels through space in the form of photons (packets of energy). Photons travel at the speed of light.
- All forms of electromagnetic radiation (EMR) have a specific frequency and wavelength that are inversely related. All EMR interacts with matter.
- The energy of a photon is directly related to its frequency.
- The energy of a photon is expressed in electron volts (eV).
- Higher energy (high frequency) photons, such as cosmic rays, may equal 10^{21}Hz. Energy may equal millions of eV.
- Lower energy (low frequency) photons may equal 60 Hz. Energy is then only a fraction of an eV.

Ionizing Compared to Non-ionizing Radiation

- Ionizing radiation is capable of dislodging orbital electrons (*e*). Photons with high energy can ionize atoms by interacting with electrons. Removing an *e* from an atom requires about 34 eV. Photons with energy >34 eV are considered ionizing.
- Non-ionizing radiation is incapable of dislodging orbital electrons, but may leave the atom in an "excited state." All lower energy (frequency) radiation is non-ionizing. These have photon energies less than 10-12 eV. This is insufficient energy to cause ionization. Primary health concerns with lower energy radiation are thermal.
- Non-ionizing radiation is expressed as a relationship of frequency, wavelength, and the speed of light. Energy = frequency x h (where h = Planck's constant, 6.63 x 10^{-34} joules/second). The higher the frequency, the higher the energy.

Frequency and Wavelength Relationship

- Frequency and wavelength are related by the equation:

$$f = \frac{c}{\lambda} \qquad \textbf{or} \qquad c = f\lambda$$

where:

c = speed of light (3×10^{10} cm/sec)
f = frequency (cycles/second)
λ = wavelength (cm)

Forms of Non-ionizing Radiation

- Forms of non-ionizing radiation include:
  - Extremely low frequency (ELF)—e.g., electricity.
  - Very low frequency (VLF)—e.g., AM radio.
  - Very high frequency (VHF)—e.g., TV, FM Radio.
  - Ultra high frequency (UHF)—e.g., TV, cellular phones.
  - Microwaves.
  - Infrared (IR).
  - Laser light.
  - Visible light.
  - Ultraviolet light (UV).

Frequency and Wavelength

- The higher the frequency, the shorter the wavelength. Conversely, the slower the frequency, the longer the wavelength.
- Non-ionizing radiation includes a wide range of frequencies and wavelengths:
  - 60 Hertz = 5×10^6 meters.
  - 300 kiloHz = 1,000 meters.
  - 30 MegaHz = 10 meters.
  - 100 MHz = 3 meters.
  - 300 MHz = 1 meter.
  - 1 GigaHz = 30 centimeters.

Ultraviolet Radiation

ULTRAVIOLET RADIATION (UV)

- Sources: The primary sources of ultraviolet (UV) light are sunlight, welding and plasma torches, and mercury vapor lamps.
- Wavelength: 100 nanometers to 400 nanometers. The ozone layer only allows wavelengths greater than 290 nanometers to reach the earth.

- Exposure Standards: Exposure standards for UV are wavelength dependent.
  - UV-A: 1 mW/cm^2 for 10^3 seconds measuring UV-A at the source.
  - UV-B & C: wavelength dependent on action spectrum, most active at 200 nm.

- Protection: Sunglasses, clothing, sunblock, and enclosing the source.

- Detection: Spectrophotometry and UV-Actinic Radiometer. These devices yield an instant reading in mW/cm^2, then compare to standard.

Fluorescent Lights : A UV Source

- An important application of UV is production of visible light from fluorescent lamps.
  - Fluorescent lamps contain a small amount of mercury vapor in a phosphor-coated tube.
  - Electrical discharge through the mercury vapor generates UV radiation that is absorbed by phosphor.
  - Phosphor fluoresces, converting short wave UV into longer wave visible.
  - Its glass tube and coating essentially absorb UV.

BIOLOGICAL EFFECTS OF UV

- The biological effects of UV are wavelength dependent. The eyes and skin are the primary organs of concern.

- Various regions of the UV spectrum cause different biological effects.

- UV-A (320-400 nm)—also termed Near UV—causes skin pigmentation changes (suntan). Chronic exposures cause premature skin aging and skin cancer. UV-A has more depth penetration than UV-B or UV-C.

- UV-B (280-320 nm)—also called Mid UV—causes erythema or skin reddening in just a few hours of exposure. Photokeratitis (sandy eye) is inflammation of the cornea with tiny lesions, and has a latency period from 0.5 - 24 hours.

- UV-C (100-280 nm)—also redered to as Far UV—has bactericidal and germicidal properties. These frequencies are found in germicidal lamps and some welding arcs.

- Chronic effects, particularly from UV-B and UV-C, are cataracts.

Various Wavelengths **Adverse Effects**

UV - A	Near UV	320 - 400 nm	Lens
UV - B	Mid UV	280 - 320 nm	Cornea
UV - C	Far UV	100 - 280 nm	Cornea

Visible Light

VISIBLE LIGHT

- Wavelength: 400 nanometers to 700 nanometers for visible light.
- Sources: Lasers, compact arc lamps, quartz-iodide-tungsten lamps, gas and vapor discharge tubes, and flash lamps.
- Exposure Standards: Visible light exposure standards are outlined in the ACGIH TLV Booklet. They depend on wavelength and exposure duration.
- Protection: Enclosure of source and interlock, protective eyewear (correctly tinted lenses).
- Detection:
  - Photon devices—photodiodes, photoconduction, and photovoltaic devices.
  - Thermal devices—calorimeters, bolometers, thermocouples.

BIOLOGICAL EFFECTS OF VISIBLE LIGHT

- Organs of greatest concern for visible light are the eyes and surrounding tissue. For many bright light sources, the normal aversion (or "blink reflex") to bright lights provides adequate protection.
- Ultraviolet light and infrared frequently occur with visible light and are frequently the biological concern.
- Acute biological effects of visible light include:
  - Retinal burns.
  - Danger of retinal damage peaks in blue light range (425-450 nm).
  - Protein denaturization.
  - Enzyme inactivation microexplosive phenomena (water on retina boils and explodes, causing scars).
  - Eye strain—tearing, blurring and headaches (short term).
- Chronic biological effects of visible light include:
  - Degeneration of the cones (color and daylight perceptors) of the retina.
  - Accelerated retinal aging (cones are destroyed over a long period of time).

Lasers

LASER SOURCES, USES, AND HAZARDS

- LASER is an acronym for Light Amplification by Stimulated Emission of Radiation.
- Stimulated Emission: The photon of one atom can cause an excited electron of a neighboring atom to drop to the same energy level, thus causing the emission of another identical photon.
- Absorption: Electrons absorb energy through optical pumping or electrical discharge. Optical pumping involves using a high-energy optical source to excite the electrons. Electrical discharge involves the use of an electrical energy source to cause collision between electrons.
- Spontaneous Emission: Electrons absorb energy from light sources such as the sun. In the absence of the light source, electrons return to their ground state, emitting light. A common example of this principle occurs in phosphorescent materials (glow in the dark).
- Absorption and stimulated emission of radiation characterize modern lasers.
- Lasers are potential eye and skin hazards.
- Lasers produce high intensity light radiation of a single wavelength.
- Lasers are not limited to frequencies of visible light.
- A carbon dioxide (CO_2) laser creates an intensely hot, but invisible, beam of infrared radiation (IR).
- Other lasers operate in the UV frequencies.
- The inverse square law does not apply to lasers. The single frequency coherent source strength of laser light does not decrease according to the inverse square law.

LASER THEORY—THE UNIQUE PROPERTIES OF LASER LIGHT

- Argon lasers may emit a pure green light, compared to an everyday light bulb, the light from which is comprised of several different frequencies. This is how laser light varies from other light forms.
- Terms used to identify laser light, its frequency, and wavelength include monochromaticity, coherence, divergence, and brightness.

- Monochromaticity: Stimulated photons are of the exact same frequency and energy level, creating pure light of one color.
- Coherence: Relates to the constancy of the light wave as it moves through space. It maintains a constant phase.
- Divergence: The degree to which a beam of light expands as it travels through space.
- Brightness: Laser light is unique, in the sense of the enormous intensity of the beam. Intensity refers to the beam power as measured cross-sectionally. The power of a laser beam is typically expressed in terms of its irradiance (watts/cm^2).

Laser Components

- Laser media: The substance (solid, liquid, and gas) that is used to create stimulated emission.
- Source of excitation energy: The method used (electrical optical pumping) to excite the laser media election.
- Totally reflecting mirror: Directs all emitted photons back through the laser media.
- Partially reflecting mirror: Also known as the "output coupler," it allows photons that have reached the desired energy levels to pass through, thus emitting the laser beam.

Laser Reflections

- Specular reflection: A laser beam strikes a reflective object and is reflected intact, basically unchanged (reflection on a mirror, belt buckle or ring).
- Diffuse reflection: A laser beam strikes a surface that reflects the beam; however, the beam is dispersed in several directions.
- Intrabeam viewing: The most hazardous form of viewing a laser beam. This is when the eye is in direct line with the beam. Specular reflections may be as hazardous and some diffuse reflections may be hazardous if the beam is refocused in the eye.

Laser Types

- There are generally two types of lasers: those that generate a pulsed wave and those that generate a continuous wave.

- Continuous wave: Releases energy at a constant rate and operates at a constant power level.
- Pulsed wave: Laser energy is delivered/released in durations of 10^{-3} seconds to 10^{-6} seconds. Pulsed lasers emit short bursts as peak power is obtained.

Types of Lasers and Laser Applications

Laser Media	Wavelength (micrometers)	Major Applications
Ruby	0.6943	Satellite ranging and medical.
Neodymium YAG	1.06	Metal processing, surgery, and range finding.
Carbon dioxide	10.6	Metal treatment and non-processing surgery.
Argon	0.488 - 0.514	Photoetching and entertainment (laser light shows and holography).
Helium-Neon	0.6328	Grocery check-out, videodisk playback, and construction.

LASER CLASSIFICATIONS

- The American National Standards Institute (ANSI) has outlined safe work practices for lasers and laser use.
- ANSI has classified lasers into specific categories. The categories range from I to IV. Class I is less hazardous than Class IV.

Class I

- Class I lasers are exempt from safe work practices. These lasers do not present eye hazards and are considered safe by all standards.
- Examples are the lasers in laser disc players and laser printers.
- The power of these lasers is usually less than 0.4 microwatts.

Class II

- Class II lasers are low power visible lasers, not capable of causing injury within the duration of the blink reflex.
- The power ranges from 0.4 microwatts to 1 milliwatt.
- Class IIa lasers are visible lasers "not intended for viewing." These are considered safe under normal circumstances. Prolonged viewing (staring) may be hazardous.
- Examples include the lasers in supermarket check-out scanners.

Class III

- Class IIIa & IIIb are medium power lasers. There is the potential for eye injury from intrabeam viewing.
- The power of class III lasers ranges from 1 milliwatt to 500 milliwatts.
- Class IIIa lasers will not produce a hazard with the blink reflexes (1–5 mw/cm^2). Examples include laser light show equipment.
- Class IIIb may produce serious eye injury from extra-beam viewing. There are no hazards from diffuse reflections (up to 500 mw/cm^2). Examples include laboratory lasers.
- Class III lasers are usually contained to reduce the classification to Class I.

Class IV

- Class IV lasers are high power lasers, usually greater than 500 milliwatts.
- Class IV lasers present eye and skin hazards and are unsafe for viewing under all conditions.
- This class of laser is used in metal cutting and welding.

BIOLOGICAL EFFECTS OF LASERS

- Biological effects of lasers are dependent on the wavelength of the laser, beam characteristics, and how the beam is viewed (i.e., direct versus reflection).
- Adverse eye effects vary with the wavelength:
  - Visible and near infrared lasers (400-1400 nanometers) cause retinal effects and damage to fovea.
  - Near UV (320-400 nm) and middle-infrared lasers (1.43-3 micrometers (μm) trigger corneal and lens effects.
  - Mid UV (280-320nm) and far infrared lasers (3μm-1mm) trigger corneal effects.
- Skin effects are usually secondary and occur as thermal damage (e.g., burns).

LASER PROTECTION

- The type of laser protection required depends on the laser's classification.
- Separate, clearly marked rooms or enclosures are necessary.

- Class I and II lasers are not posted.
- Class IIIa and IIIb lasers require "Caution" signs and well-lighted areas to decrease pupil size.
- Class IV requires "Danger" signs.
- User training is required.
- Proper protective eyewear must be selected and used. Consider the following when selecting protective eyewear:
  - Wavelength.
  - Radiant exposure.
  - Maximum permissible exposure.
  - Optical density provided.
  - Visible light transmission required.
  - The need for prescription eyewear.
- Direct viewing and specular reflections must be prevented.
- Other associated hazards include:
  - High voltage: More injuries from using lasers occur from electrocution than the laser beam itself.
  - Compressed gases: Some lasers use compressed gases as laser media or for beam enhancement. Welding and cutting exotic metals may need local exhaust ventilation.
  - Fire: Unexpected beam reflections can ignite flammable and combustible materials in their path.
- Class IV lasers can be totally enclosed and interlocked, preventing exposure. The laser "system" could then be reclassified as a Class I system with an embedded Class IV laser. However, during maintenance, it reverts back to Class IV.

Biological Monitoring

- Most laser user examinations include:
  - Fundus photography (photograph of retina).
  - Visual activity.
  - Visual fields and depth perception.
- Exams are generally provided at a minimum upon initial assignment to laser use and upon termination of laser use.

Optical Density (OD)

- Optical density is a term used to define the protection level of protective eyewear for non-ionizing radiation. The optical density for welding goggles may be 14. A pair of specific protective laser goggles may have an OD of 7.

- Half value layers determine the relationship between the absorption of monochromatic visible radiation and the length of the path through the absorbing medium. The symbols I_o and I represent incident and transmitted beam intensity, respectively. Log (I_o/I) represents optical density or absorbance.

- The formula for calculating optical density is outlined below.

$$OD = \text{Log}\,(I_o/I)$$

where:

OD = optical density
I_o or I_1 = initial beam intensity
I or I_2 = final beam intensity

Sample Question

- What is the optical density of a pair of laser safety glasses that reduces beam intensity from 500 milliwatts/cm^2 to 5 microwatts/cm^2?

$$OD = \text{Log}\,(I_1/I_2)$$

$$\left(\frac{I_1}{I_2}\right) = \text{Log}_{10}\left(\frac{500\text{mw}/\text{cm}^2}{5\mu\text{w}/\text{cm}^2}\right) = \text{Log}_{10}\left(\frac{500\times10^{-3}\,\text{w}/\text{cm}^2}{5\times10^{-6}\,\text{w}/\text{cm}^2}\right)$$

$$= \text{Log}_{10}\,(10^5) \cong 5$$

Infrared Radiation

INFRARED RADIATION (IR)

- All objects emit infrared radiation to other objects that have lower surface temperature.

- If not controlled, infrared radiation may cause skin burns, cataracts (heat cataract or glass blower's cataract), and retinal damage.

- IR can cause sweating, in the event of heat overload. Heat exhaustion or heat stroke can occur.

- Wavelength: 700 nanometers to 1 millimeter.

- Sources: Infrared heaters, glass blowing, blast furnaces. Infrared light is heat. Anytime you are exposed to temperatures warmer than you are, infrared occurs.

- Exposure standards: Exposure standards can be found in the ACGIH TLV Booklet. To use this information, the wavelength, geometry of source, and length of exposure must be known. For example: 10 milliwatt/cm^2 for wavelengths > 770nm (reference: 1 milliwatt/cm^2 = summer day at noon).

- IR lasers: Separate TLVs are based on wavelength and exposure duration.

- Protection: Enclosures, protective eyewear (tinted lenses or laser goggles) may be required.

- Detection:
  - Spectrophotometers—light output at each wavelength.
  - Calorimetry—measures temperature increase.

BIOLOGICAL EFFECTS OF IR

- Biological effects of infrared radiation are wavelength dependent.

- Infrared radiation may cause cataracts and flash burns of the skin and cornea, similar to radio frequency and microwave radiation.

- The blink reflex appears to provide protection against acute exposure, but protection measures are required for chronic exposures.

		Wavelength	**Penetration**
IR - A	Near IR	700 - 1400 nm	Deep
IR - B	Mid IR	1400 - 3000 nm	Shallow
IR - C	Far IR	3000nm - 1mm	Superficial

Microwave/Radio Frequency Radiation

MICROWAVE/ RADIO FREQUENCY (RF) RADIATION

- Microwave and radio frequency radiation includes Frequencies ranging from 0.1cm to 300 meters or 1MHz to 300,000 MHz.

- Sources: Antennas, radar transmitters, microwave ovens, diathermy (deep heat therapy), and RF sealers on induction heaters.

- Microwaves create heat by causing water molecules to vibrate, get agitated, and heat up.

- Microwaves are reflected by metal. They pass through glass, paper, and plastic. Materials containing water absorb them.

- Wavelengths: The greatest potential for radiation to penetrate the body and damage internal organs is between 0.25 and 2 meters, because it does not set off the heat receptors in the skin.

- Exposure Standards: TLVs vary with wavelength.

Microwave and Convection Ovens

- Convection (traditional) oven: Food must accept the heat on the surface and then conduct the heat inside.

- Microwave Oven: Radiated microwaves are absorbed throughout the medium. Materials containing water readily absorb microwaves. Most microwave ovens operate at either 915 or 2,450 MHz. Wavelength is between 0.1–1 meter.

Microwave Exposure Standards—300 MHz to 300 GHz

- Exposure standards for microwaves are based on thermal effects.

- Microwave oven standards:
  - Units must emit < 1 mW/cm^2 at any point 5 cm or more from the oven prior to purchase.
  - Units may emit < 5 mW/cm^2 at any point 5 cm or more from the oven after purchase (measured around seal of oven door).

- Workplace exposure standards for other RF sources are based on thermal effects:
  - OSHA: 10 mW/cm^2 continuous exposure.
  - ACGIH: Frequency dependent. Specific absorption rate (SAR) varies by frequency.

Microwave Detection

- Microwave Oven Survey Meter: Broadband RF meters will not disturb the field being measured. Broadband RF has a probe to separate electric and magnetic field measurements.

Near Field and Far Field

- Near Field: Relationship between electric field (e) and magnetic field (h) is not constant, so both must be measured. Separate measurements of the electric field and magnetic field strengths must be made and compared with standards.

- Far Field: There is a constant relationship between the fields. Power density in mW/cm^2 is directly measurable. The inverse square law is applicable to far field. There is a constant relationship between the electric field and magnetic fields (plane wave). Power density in $milliwatts/cm^2$ is directly measurable.

Determining Power Density

- The formula below is used to determine the power density of the far field, given distance, antenna area, wavelength, and power output.

$$W = \frac{AP}{\lambda^2 r^2}$$

where:

W = power density in mW/cm^2
A = effective antenna area in cm^2
P = average power output in mW
λ = wavelength in cm
r = distance from antenna in cm

BIOLOGICAL EFFECTS OF MICROWAVE AND RF

- The most sensitive organs for microwave and RF exposure are testes and eyes. In testes, the effect can be temporary sterility.
- In the eye, the concern is generally with the lens. The lens can't handle heat because there is no blood supply to carry away heat. Cataracts are often a concern of non-ionizing radiation due to hardening of the cornea.
- Primary health concerns are thermal: burns, cataracts, effects on reproduction, and death. Environmental temperature can impact the ability to handle the heat generated by microwave energy. The higher the frequency, the lower the health hazard because the depth of penetration is less.
  - >10,000 MHz—skin absorption.
  - 1000-10,000 MHz—deeper penetrations, cataract formations.
  - 1-1000 MHz—deepest penetration. Damage to internal organs, lens of eye, and testes.

Extremely Low Frequency Radiation

EXTREMELY LOW FREQUENCIES (ELF)

- The most prominent source in the extremely low frequency (ELF) region is 60 Hz electric power. ELF is at the very low end of the EM spectrum, therefore the energy is very low.
- Wavelength: 50-60 Hz.
- Exposure standards: ACGIH standards are based on understood, verifiable health effects (e.g., magnetophosphenes, induced currents, and potential interference with electronic devices, like pacemakers).

- Research into potential impacts, such as cancer (e.g., leukemia, brain tumors), continues, but currently no claims are considered scientifically substantiated.

Electric Fields and Magnetic Fields

- With a wavelength of 5×10^6 meters, the near field is always present. Therefore, both *e* and *h* fields must be measured separately.
- The *e* field is related to voltage or electric field. The most obvious hazards relate to shocks and currents from high voltage lines and equipment.
- The *h* field is related to the current or magnetic field. These low frequency magnetic fields mostly generate from natural sources, such as thunderstorms and solar activity.
- Most interest into potential health effects is focused on magnetic fields generated from synthetic sources (<0.3 microTesla (μT)). Sources include home and office video display terminals (VDTs).
- Protection from electric fields:
  - Shielding—At virtually any ELF frequency, a conductive surface will provide substantial electric field shielding.
  - Access to areas where electric field strengths are high should be limited.
- Protection from magnetic fields:
  - There is no practical way to shield against ELF magnetic field exposure in all cases.
  - Exposure to areas where magnetic fields are strong should be limited.

BIOLOGICAL EFFECTS OF EXTREMELY LOW FREQUENCIES

- ELF electric fields externally applied to surfaces of the body induce electric currents and fields inside the body and excite cells.
- ELF magnetic fields do not cause perturbation.
- ELF's central nervous system effects are primarily on circadian timing mechanisms.
- Unconfirmed data: Most human studies conducted are based upon surveys of utility workers and people living in the vicinity of high-voltage lines.
- There is insufficient energy to cause direct genetic damage, as well as insufficient energy to cause heating.

11 Ergonomics and Human Factors

The human body can endure significant discomfort. The discomfort can originate from many awkward and unnatural movements, or environmental factors such as poor lighting, poor indoor air quality, and thermal stress. An industrial hygienist must be knowledgeable in these areas to effectively evaluate their impact on workers and recommend appropriate control measures to minimize this discomfort.

Ergonomics

ERGONOMICS

- ❑ Ergonomics considers the total physiological and psychological demands of a worker's job, rather than just productivity, health, and safety.
- ❑ The benefits of this approach minimize the physical stress on workers by reducing the incidence of repetitive motion injuries, reducing injury rates, making operations more efficient, lowering costs of operation, reducing training time, and more effectively using human resources.
- ❑ The human body can endure significant discomfort and stress. However, when awkward conditions or motions are continued for prolonged periods, they can exceed a worker's physiological limitations.
- ❑ Consider biomechanics, workplace design, and behavioral components of a job when reviewing ergonomic related issues.

Impact of Ergonomic Problems in the Workplace

- ❑ The ergonomic problems in today's workplace require understanding the impact on employees and employers. These include:
  - ❑ Pain and loss of function in soft tissues.
  - ❑ Worker fatigue.
  - ❑ Poor work performance.
  - ❑ Reduction in quality and productivity.
  - ❑ Increase in worker's compensation claims.
  - ❑ Increased insurance cost for company.

Prevention Responsibility

- ❑ Prevention responsibility is shared by the employer and employee.
  - ❑ An employer has the responsibility to provide a safe workplace and to require safe work practices.
  - ❑ An employee has the responsibility to follow safe work practices.
- ❑ When top management of an organization understands how many dollars are wasted on injuries, treatments, and workers' compensation, support is usually provided.

- An organization's safety and occupational health department should seek out risk factors, then reduce and eliminate them.

Ergonomics Training

- A successful employer approach is to ask, "What can we (as a company) do to make you less tired when you leave work?" This approach has been very successful at companies such as Steelcase.

- Other types of training have not been as successful. NIOSH has shown that individuals who received "lifting training" have just as many injuries as those who have received no such training.

Company Ergonomic Programs

- Programs should address and quantify improvements in:
  - Worker fatigue.
  - Poor work performance.
  - Discomfort and pain.
  - Employee relations.

Challenges to an Ergonomics Program

- There are many challenges to an ergonomic program. The challenges include rigid job codes and unions. A major factor to remember is that employee lifestyles (such as drinking, smoking, and lack of exercise) will affect worker performance.

- There may be difficulty in showing the financial benefits of an ergonomic program, and the current business climate may adversely affect a program (e.g., asking for more from fewer workers).

- Other challenges to an effective ergonomic program include worker resistance to changes in work habits, limited time and resources to adequately train workers, and the ability to justify purchases of new tools when downsizing has resulted in a surplus of the poorly designed type.

Quality, Productivity and Ergonomics

- The most successful ergonomic programs can directly relate ergonomic improvements to:
  - Improved quality of the product.
  - Increased productivity.
  - Decreased OSHA recordable injuries (or medical visits).
  - Fewer worker compensation claims.

CUMULATIVE TRAUMA DISORDERS (CTDs)

- Cumulative trauma disorders (CTDs) are disorders of the musculo-tendinous-ossesous-nervous system that are caused, precipitated, or aggravated by repeated exertions or movements of the body. This definition reflects the philosophy of workers' compensation laws in most western countries.

Cumulative Trauma Disorders (CTDs) — Generalization

- Cumulative trauma disorders can be generalized as disorders of the musculo-skeletal system, produced by tiny, but cumulative, amounts of damage.
- The damage occurs on a daily basis and can be caused by repetitive motions or sustained posture.

Characteristics of Cumulative Trauma Disorders

- Common characteristics of cumulative trauma disorders include:
  - Involvement of mechanical and physiological processes.
  - Illness relates to work intensity and duration.
  - Development of trauma requires weeks, months, and years.
  - Recovery requires weeks, months, and years.
  - Symptoms are often poorly localized, nonspecific, and episodic.
  - Symptoms are often multifactorial (i.e., more than one causal factor).

Physiological Processes of CTD in the Musculo-Skeletal System

- The following summarizes the physiological processes of the musculo-skeletal system.
  - Muscles attach to tendons.
  - Tendons attach to bones.
  - Moving bones are accommodated at joints.
  - Joints are held together by ligaments.
  - Circulation occurs through joints by blood vesels.
  - All are controlled by a complex nervous system.

Examples of Cumulative Trauma Disorders

- CTD is not a diagnosis. It is a group of disorders with similar characteristics. Some examples of CTDs include:
  - Myalgia—pain in one or more muscles.
  - Tendonitis—inflammation of a tendon.
  - Tenosynovitis—inflammation of the connective tissue sheath of a tendon.
  - Epicondylitis—inflammation and pain of certain bony prominences in the area of the elbow, usually resulting from excessive strain on and twisting of the forearm (for example, tennis elbow).
  - Trigger finger—sometimes called snapping finger, a condition of partial obstruction in flexion or extension of a finger. Beyond the point of obstruction, movement is eased. Caused by constriction of a tendon sheath.
  - Carpal tunnel syndrome—a common affliction caused by the compression of the median nerve in the carpal tunnel. Often associated with tingling, pain, or numbness in the thumb and first three fingers.
  - Raynaud's syndrome—abnormal constriction of the blood vessels of the fingers on exposure to cold temperature, usually caused by low frequency vibrating hand tools.

Raynaud's Syndrome

- ❑ Raynaud's syndrome is sometimes called "vibratory white finger," or "white fingers." The fingers turn white because there is less blood flow to the digits.
- ❑ Raynaud's syndrome is caused by vibrations in the 500 to 1200 Hz frequency range, usually transmitted to the hands and fingers by vibrating hand tools.
- ❑ Exposure to cold aggravates this condition. Anti-vibration gloves help. It sometimes occurs in newer workers who may grip too tightly.

CTD Individual Risk Factors

- ❑ Individual risk factors play a significant role in the development of cumulative trauma disorders. These include:
  - ❑ A lack of joint mobility and muscle flexibility due to arthritis or other abnormal body and joint functions.
  - ❑ Cysts, swellings, and other abnormalities of tendons.
  - ❑ Poor circulation that impairs repair mechanisms.
  - ❑ Fluid retention from circulatory problems, high blood pressure, pregnancy, salt imbalance, thyroid imbalance, etc.
  - ❑ Poor dietary habits (vitamin B-6 is important to the repair mechanism of tissues).
  - ❑ Alcohol and tobacco use affects the musculo-skeletal system.
  - ❑ Nicotine is thought to break down connective tissue fibers.
  - ❑ Diseases such as arthritis affect joints.
  - ❑ Diabetes can reduce circulation.
  - ❑ Previous injuries can distort function and increase irritability of injured tissues.
  - ❑ Hormone shifts can occur in women.
  - ❑ Emotional stress tightens blood vessels.

Localized Fatigue: A First Indicator of CTD

- ❑ Fatigue is one of the first symptoms of cumulative trauma. This fatigue can occur in any muscle group from sustained postures, excessive motions, awkward movements, etc.
- ❑ Characteristics of fatigue include:
  - ❑ Involvement of both mechanical and physiological processes.
  - ❑ Results in pain and impairs work performance.
  - ❑ Related to the intensity and duration of work.
  - ❑ Occurs within seconds, minutes, or hours.

Daily Wear and Tear and it's Relationship to CTD

- ❑ If the injury rate of the structures of the body is greater than the repair rate, cumulative trauma begins to occur. CTD occurs when the work involves repetitive stresses over time.

- The "selection of fittest rule," or the "healthy worker effect," applies in many highly repetitive, strenuous jobs. This means that many less physically fit individual employees willingly leave or quit jobs that cause pain and discomfort.

- The destruction process occurs as follows:
  - Contractions produce pressure.
  - Pressures can reduce circulation.
  - Individual muscle fibers can tear.
  - Joint surfaces rub together under pressure.

- There is a tremendous reserve of tissue in the human body, so that large amounts of damage can be covered or hidden for some time before the onset of fatigue, pain, lost time injury, and disability.

Injuries

- Injury to the muscle groups can be from repetitive motions or sustained postures.

- Effects of a sustained posture can be just as dangerous as repetitive motions, because:
  - Sustained postures burn tremendous amounts of energy.
  - Muscles reset themselves tighter and tighter.
  - Sustained contractions of neck and muscles can compress blood vessels and nerves.
  - Some of these blood vessels and nerves supply arms with impaired circulation and, therefore, less repair occurs.
  - With nerve pressure comes pain, tingling, and numbness.

PREVENTION AND REPAIR

- Preventing ergonomically related illness or injury is the goal of industrial hygienists. Reducing the rate of damage and increasing the rate of repair can allow this.

- Prevention and repair can be achieved through:
  - Good nutrition.
  - Good circulation.
  - Adequate rest.
  - Walking or movement at breaks.
  - Maintaining neutral postures.
  - Medical treatment for conditions that contribute to the problem.
  - Stress reduction.

Reducing the Rate of Damage

- Reducing the rate of damage helps prevent ergonomically related disorders. This reduction can be accomplished by a variety of methods, including:
  - Eliminating awkward, sustained postures.
  - Varying the work to avoid repetitive motion of limited muscle groups.
  - Avoiding excessive repetitive reaches.
  - Avoiding excessive repetitive force.
  - Alternating postures and movements.
  - Limiting frequency and weight of manually lifted objects.

Jobs at High Risk for CTD

- Jobs at high risk for cumulative trauma disorders or other ergonomically related illnesses or injuries include jobs with the following characteristics:
  - Low effort risks.
  - Highly focused and limited to a small number of movements.
  - Performed all day by a small number of muscles and tendons.
  - Using repetitive motions and unchanging body postures.

Selecting Ergonomic Improvements

- Identifying and evaluating an ergonomic problem takes time. Seek information from as many affected workers as possible. Then study literature on suspected causes of CTDs.

- Test tools, furniture, job changes, etc., in a pilot area before purchasing expensive products or changing an entire work area.

- Evaluate effectiveness after the implementation of these solutions.

Biomechanic Principles

- Biomechanics deals with the structural elements of the body and the effects of the external and internal forces on various parts of the body.

- Review flexion and extension positions of the hand, elbow joints, and shoulders while work is performed. Flexing motions of the joints repeated rapidly over a long period of time result in inflammation of the tendon sheaths, which may cause tendonitis, trigger finger, carpal tunnel syndrome, epicondylitis, and thoracic outlet syndrome, among other illnesses. Joints in neutral positions result in less stretching of soft tissues and therefore, less stress.

Biomechanic Control Methods for CTD

- Control methods for preventing ergonomically related disorders include a variety of techniques.

- Keep work close to the body to minimize horizontal leverage.

- Minimize prolonged forward bending. This puts stress on muscles and ligaments of the back.

- Eliminate twisting the trunk. This stretches the elastic disc between vertebrae and stresses muscles on both sides of the spine.

- Eliminate sudden forceful movements. Sudden forceful movement places excessive stress on muscles. Movements should be even and gradual.

- Alternate postures as well as movements to reduce fatigue from poor circulation. Use a variety of muscle groups, particularly large and small muscle groups.

- Limit duration of any continuous muscular effort to allow for recovery and prevent muscular exhaustion that has a much longer recovery period.

- Distribute resting time. Frequent shorter breaks are more beneficial than fewer longer breaks.

Physiological Factors and Control Methods

- Physiological factors include energy expenditure during lifting, working, standing, etc.

- Control measures include limiting energy expenditure in a task. The amount of energy that the heart and lungs can supply to the muscles is limited.

Anthropometric Factors and Control Methods

- Anthropometric factors signal the fact that all workers are not the same size.

- Workers vary in proportion as well as size.

- Use anthropometric tables when designing workstations. Some tasks require that you design for shorter workers, some for taller workers.

- Adjustability is an initial extra cost but a cost savings in the long run. Furniture companies are a good source for this information.

Workplace Design

- Workplace design considers the job, the equipment used by the workers, and a workplace survey. The survey identifies how biomechanic, physiological, and anthropometric factors work effectively to minimize human injury, increase worker efficiency, and decrease human error.

Posture: Sitting

- Prolonged sitting is stressful for the neck and back, particularly if hands have to be in visual focus. It leads to a forward leaning posture.

- ❑ Provide adjustable chair seat height, and back lumbar support. Eliminate twisting by providing a swivel mechanism.
- ❑ Provide worker with proper training, including specific chair and work surface height adjustment.
- ❑ Use the task to determine the chair type needed. The task—taking hand and arm use and visual demands into consideration—should determine the height of the work surface.
- ❑ Minimize bending and twisting of the trunk by avoiding excessive reaches.
- ❑ Select a sloping surface for reading tasks (e.g., 45° for reading only and 15° for reading and writing).
- ❑ Allow sufficient legroom.

Posture: Standing

- ❑ Alternate standing with sitting and walking. A sit-stand workplace or stool is recommended.
- ❑ Work surface height should be based on the task and worker size.
- ❑ Provide room for legs and feet. Footrests, such as in a bar, reduce stress on the back muscles.
- ❑ Avoid excessive reach to minimize leaning over and twisting trunk.

Posture: Arms and Hands

- ❑ Select the right type of tool to maintain a neutral posture.
- ❑ Use a curved tool instead of bending the hand or wrists.
- ❑ Hand-held tools must not be too heavy.
- ❑ Maintain tools.
- ❑ The shape and locations of handgrips on equipment and tools should allow for neutral positions.
- ❑ Tasks should be between waist and shoulder level.
- ❑ Tasks should allow hands to be in front of body.

Posture: Lifting, Carrying, Pulling, and Pushing

- Lifting should keep lumbar curve intact while performing a covert lift.
- Limit pulling and pushing forces required, minimizing back stress.
- Pulled and pushed loads need adequate leg and foot space underneath.
- Position vertical, cylindrical hand grips at a comfortable height.
- A trolley should have two swivel wheels on the side where the handgrips are located. Four-swivel wheels require constant steering.
- Floors should be hard and even.

Indoor Air Quality

HISTORY AND CAUSES OF POOR INDOOR AIR QUALITY (IAQ)

- Many indoor air quality problems originated with the fuel shortages in the 1970s and as energy efficient homes and buildings were built in the late 1970s and 1980s.
- The ability to open windows has disappeared in most commercial buildings. Control of the workplace environment airflow has become mechanically controlled.
- Many alternative building materials were used during the 1970s and 1980s to conserve energy and other natural resources.
- The sensitivity of the public to indoor air quality standards has increased due to publicity and historical incidents such as the infamous Legionnaires disease event.

Useful Terms for Understanding Indoor Air Quality Issues

- Air Handling Unit (AHU): Ventilation equipment in the HVAC system.
- American Society of Heating, Refrigeration, and Air Conditioning Engineers (ASHRAE): Primary North American association dealing with IAQ issues, which has developed a number of IAQ-related standards.
- Heating, ventilating, and air conditioning (HVAC): Air-handling systems designed primarily for temperature, humidity, odor, and air quality control.
- Outdoor Air (OA): Fresh air or outdoor air that is mixed with return air (RA) to dilute contaminants in the supply air (SA). Outdoor air is usually obtained from outside the building, but alternatives exist, such as from an acceptable hallway.

Building Design and Maintenance

- Building design and maintenance factors can influence the occurrence of indoor air quality:
- Building for occupancy number and function.
- Building for energy efficiency.
- ASHRAE standards.
- Ventilation systems, design, balance, and maintenance.
- Whether smoking is permitted.

HEATING, VENTILATION, AND AIR CONDITIONING (HVAC) SYSTEMS

- Mechanically ventilating, heating, cooling, humidifying, dehumidifying, and cleaning air for comfort, safety, and health is done through HVAC systems.

- HVAC systems also provide odor control, and they maintain oxygen and carbon dioxide levels at acceptable concentrations.

- HVAC systems are built to serve specific zones. Most zones are controlled by one thermostat.

HVAC Operation Simplified

- The simplified operation of a heating, ventilating, and air conditioning system is outlined below:
  - Outside air (OA) enters through an intake, pushed by atmospheric pressure. A damper regulates the amount of OA taken into the system.

  - Incoming OA mixes with return air (RA) from the occupied space, forming recirculated air, or mixed air (MA). The MA passes through filters and then enters the fan.

  - Discharged at the fan outlet, the air, now under positive pressure, is pushed through coils that heat and cool, depending upon need. A drain pan is fitted below the coils to collect water that condenses on the cooling coil.

  - The air moves through metal ductwork (which must be kept dry and mold free) at 1000-2000 feet per minute (fpm) to a distribution box. From there, it travels through smaller ducts to the supply terminal or diffusers.

  - Air enters rooms at 500-1000 fpm and slows to a terminal velocity of about 4-50 fpm.

  - The air migrates through the space to the return air register, to the return air duct, where it may be recirculated or exhausted (EA).

Constant Volume Systems

- A constant air volume (CV) supply system varies the air temperature to maintain room temperature, while a variable air volume (VAV) system varies the amount of air delivered to the space to maintain temperature.
- Example: In summer, if the temperature increases, additional cool air is provided. As the space temperature drops to desired levels, the airflow diminishes.

Variable Volume Systems

- A variable air volume (VAV) system varies the amount of air delivered to the space to maintain temperature.
- Variable volume systems are energy efficient, and have low operating costs. Air volume and fan speeds are linearly related. If the airflow is to be cut in half, then the fan speed is cut in half. Horsepower and speed are related through a third-power relationship. If the speed is reduced by half, the horsepower is reduced eight times.
- A disadvantage of variable volume systems is that they may not provide sufficient outside air.

Air Mixing

- Delivery of air to a room does not guarantee mixing. Offices with partitions are susceptible to poor air mixing. Mixing can be quantified by the formula for mixing efficiency or ventilation effectiveness.
- The mixing factor is defined as the ratio of the amount of air required to dilute a contaminant to the ideal amount of air that should be used to reduce it. Use the formula below when quantifying mixing efficiency or ventilation effectiveness.

$$K_m = \frac{Q_{Actual}}{Q_{Ideal}}$$

where:

K_m = mixing factor
Q_{Actual} = volume of air actually needed to dilute contaminant
Q_{Ideal} = volume of air ideally needed to dilute contaminant

- In well-planned buildings served by good HVAC systems, K_m is often close to 1.0 (perfect mixing). Good mixing can be attained by:
  - Providing and properly positioning supply and return registers.
  - Placing supply registers to circulate air where people are located (in the occupied zone).
  - Supplementing poor zones with free-standing fans.

HVAC Sources and Causes of IAQ Problems

- Fifty percent of IAQ problems are related to deficiencies in the ventilation of the building.
  - Lack of outside air.
  - Poor air distribution.
  - Uncomfortable air temperatures and humidities.
  - Sources of contaminants in the air ventilation system.

- Thirty percent of IAQ problems related to some indoor contaminant, including:
  - Formaldehyde.
  - Solvent vapors.
  - Dusts.
  - Microbiological agents.

- Ten percent of IAQ problems are attributed to an outdoor source, including:
  - Motor vehicle exhaust.
  - Pollen.
  - Fungi.
  - Smoke.
  - Construction dust.

- Ten percent of IAQ problems have no observable cause.

HVAC Standards for Maintaining IAQ

- ASHRAE 62 (1989): This standard lists minimum OA requirements for ventilation of acceptable air quality for about 100 occupancies, including offices and classrooms. The standard assumes good mixing and distribution of outdoor air. It also assumes a certain occupant loading.

- ASHRAE 55 (1992): This standard specifies thermal environmental conditions for human occupancy in which 80-90 % or more of the occupants should find the environment thermally acceptable. It does not address other environmental factors, such as air quality.

- ASHRAE 52 (1992): This standard specifies methods of testing air cleaning devices used in general ventilation for removing particulates.

Regulations

- OSHA has proposed an IAQ regulation. Its major ventilation provisions include:
  - Maintaining carbon dioxide (CO_2) levels below 800 ppm. This translates into an OA requirement (in typical office spaces) of 25-30 cubic feet per minute (cfm).
  - Keeping relative humidity below 60 percent.

- Obtaining and maintaining records on HVAC systems.
- Inspecting, maintaining, and operating the HVAC system so that it meets the criteria of the codes in force at the time the building was constructed.
- Exhausting designated smoking areas to the outdoors and keeping these areas under negative pressure.
- Locating intakes to prohibit the uptake of contaminated air.
- Providing local exhaust of specific emitters, if necessary.

Typical HVAC Deficiencies

- There are a number of typical HVAC deficiencies. Insufficient air delivery, insufficient outside air, poor air distribution, and poorly maintained equipment are a few examples. Descriptions of the typical HVAC deficiencies are outlined below.

- Insufficient total air delivery to an occupied space may be caused by:
  - Inadequate fan capacity.
  - Worn fan blades.
  - An increased number of occupants.

- Insufficient OA delivered to an occupied space may be caused by:
  - OA dampers set too low.
  - Unbalanced supply and return systems.
  - Insufficient temperature control capacity.

- Inadequate air distribution within a space may be caused by:
  - Poorly operating dampers.
  - Blocked diffusers.
  - Floor resting office partitions.

- AHU components may not operate properly if the:
  - System does not start before arrival of occupants, and shuts down before departure.
  - Filters are inadequate.
  - Room temperature and humidity controls are inoperative.

Percentage of OA in the Supply Air

- Use the formula below when calculating the percentage of outside air in the supply air.

$$\%OA = \frac{T_{RA} - T_{MA}}{T_{RA} - T_{OA}} \times 100$$

where:

T_{RA} = temperature of return air (dry-bulb)
T_{MA} = temperature of mixed return and outside air (dry-bulb)
T_{OA} = temperature of outdoor air (dry-bulb)

Estimating the Volume Flow Rate of OA

- Use the constant emission formula described below to determine the volume flow rate (in cfm) of outside air.
- The number of people working in an office space and the measured carbon dioxide concentration in the space (after the concentration has leveled off) are generally related to the volume flow rate of outdoor air as follows:

$$Q_{OA} = \frac{13{,}000n}{C_{indoors} - C_{OA}}$$

where:

Q_{OA} = approximate volume flow rate of outdoor air (cfm)
n = number of people working in an office complex (with about seven people per 1,000 square feet of office space)
$C_{indoors}$ = measured concentration of CO_2 in the office air after a long period of occupancy time, such as near lunch or near the end of the day (ppm)
C_{OA} = concentration of CO_2 in the outdoor air (ppm)

Air-exchange Rate Calculation

- Tracer gases are used to assist the evaluation of an indoor air pollution problem. Outlined below is a formula often needed when using a tracer gas.
- In the concentration-decay tracer gas approach, human origin carbon dioxide concentrations are used as the tracer gas. During the day, carbon dioxide builds up naturally. After everyone leaves, OA will dilute the carbon dioxide still present. Knowing the initial and final concentrations and the time elapsed allows the use of the purge formulas to predict the amount of OA delivered.

$$N = \frac{\ln(C_i - C_o) - \ln(C_a - C_o)}{h}$$

where:

N = air exchange, air changes per hour, OA
C_i = concentration of CO_2 at start of test
C_o = outdoor concentration, about 330 ppm
C_a = concentration of CO_2 at end of test
ln = the natural log

Control Methods

- Control factors to reduce and prevent indoor air pollution include:
  - Increasing occupant awareness.
  - Understanding that most indoor air problems relate to the general ventilation system.
  - Considering the workplace environment.
  - Considering potential sources such as copiers, office machines, and computers.
  - Quality of life.

Illumination

PURPOSE OF LIGHTING

- Good lighting is necessary for:
- Ability to comfortably see visual tasks.
- A safe working environment.
- Decreasing mistakes.
- Increasing production (Hawthorn effect).
- Improving morale (Hawthorn effect).
- Reducing accidents.

Illumination Requirements for Industry

- The American National Standard Practice for Industrial Lighting contains a list of recommended levels of illumination for specific visual tasks and areas. These recommendations are based on task characteristics and the visual performance requirements of young adults with normal eyes.

- The recommendations are intended to provide guidelines desirable from an overall operational viewpoint rather than from safety alone. The recommended levels do not take into account the use of safety goggles that may reduce light reaching the eyes. Example: The table below represents an example of the information found in the ANSI standard.

Levels of Illumination Currently Recommended

Area	Foot-candles on Tasks
Clothing manufacturer (men's)	
Receiving, opening, storing, shipping	30
Examination	2000
Cutting	300*
Shops	100
Inspection	500*
Pattern making, preparation of trimming, piping, canvas, and shoulder pads	50

* Can be obtained with combination of general lighting plus specialized supplementary lights. These seeing tasks generally involve the discrimination of fine detail for long periods of time and under conditions of poor contrast. The design and installation of the combination system must provide sufficient amount of light, but also the proper direction of light, diffusion, and color and eye protection. As far as possible, it should eliminate direct and reflected glare, as well as objectionable shadows.

Industrial Lighting Equipment

- ❑ Electric light sources used for industrial applications fall into three categories:
  - ❑ Incandescent (including tungsten-halogen).
  - ❑ Fluorescent.
  - ❑ High intensity discharge (including mercury, metal halide, and high pressure sodium).

- ❑ There are two categories of luminaires for industrial lighting:
  - ❑ General.
  - ❑ Supplementary.

- ❑ General luminaires are used to provide general lighting throughout an area or localized general lighting. They are classified into five types according to their light distribution:
  - ❑ Direct lighting (most efficient, but may produce shadows and glare).
  - ❑ Semi-direct lighting.
  - ❑ General diffuse.
  - ❑ Semi-indirect lighting.
  - ❑ Indirect lighting.

- ❑ Supplementary luminaries are used along with a general lighting system but are located near the seeing task to provide the higher levels or quality of lighting not readily obtainable from the general lighting system.

LIGHTING MEASUREMENTS AND INSTRUMENTS

- ❑ Illumination (light) meters measure three essential quantities:
  - ❑ Illumination (in foot-candles): Measurements made at task locations for task lighting and at various locations along the horizontal work surface for general lighting.
  - ❑ Luminance (in footlamberts): Measurements made of luminaires and room and task surfaces.
  - ❑ Reflectance (in percentages): Measurements made of room surfaces.

- ❑ Luminance (brightness) meters measure the luminance of surfaces using an illumination meter if the reflectance of the surface is known (luminance = illumination x reflectance).

Intensity

- ❑ Intensity is how much light a source gives off in a given direction. The unit of luminous intensity is the candela (formerly candle), sometimes referred to as candlepower.

- A lumen is a unit of light output from a light source. A 100-watt incandescent lamp emits about 1700 lumens initially in all directions. A 40-watt cool-white fluorescent lamp emits about 3200 lumens initially.

- One candela = 4π lumens.

Illumination Level

- Illumination level is the quantity of light falling on a surface, measured in foot-candles. Example: If 100 lumens from a light source fall on one square foot of a table top, the illumination level on the table would be 100 foot-candles.

- If the unit of surface area is in square meters rather than square feet, the illumination is measured in lux.

Luminance

- Luminance, or photometric brightness, is a measure of light emitted or reflected from a certain area of a surface. Its unit is footlamberts (in square feet), one lumen per square foot = one footlambert. In square meters, luminance is referred to as candelas/square meter.

Reflectance

- Reflectance is how much light is reflected from a surface. It is the ratio of the luminance of a surface to the illumination on the surface.

- A completely black surface has a 0 percent reflectance. A white surface has a reflectance of 1.0 or 100 percent.

Glare

- Discomfort glare produces discomfort and may affect human performance, but it does not interfere with visual performance or disability.

- Disability glare does not cause pain, but reduces the visibility of objects to be seen (e.g., night driving with oncoming headlights).

- Both are caused by bright light sources and by bright reflections. Glare can be avoided by mounting luminaires as far above or away from normal lines of sight as possible.

Thermal Balance

THERMAL STRESS

- Thermal stress is a physical agent in many working environments. These environments may be outdoors, where air temperatures range from -20 to 110° F in different regions of the United States.

Thermal Balance Model

- Thermal balance is described with a heat balance equation, in which the major avenues of heat exchange between the body and the environment are represented by variables. (There is no uniformly accepted version of this formula.)

- Heat Balance Equation:

$$S = M \pm R \pm C - E$$

where:

S = rate of storage in the body, or oxygen requirement
M = rate of metabolism (always positive) (BTU/hr)
R = rate of radiant heat loss or gain (BTU/hr)
C = rate of convective heat loss or gain (BTU/hr)
E = rate of evaporative heat loss (always negative)

Factors Affecting Thermal Stress

- Three main factors that influence the degree of thermal stress are:
  - Climatic conditions.
  - Work demands.
  - Clothing worn.

- Climatic conditions are widely used to describe the degree of thermal stress. These include air temperature, wind chill, and humidity.

- The role of metabolic rate in heat balance is a substantial contributor to heat gain. In heat stress, metabolic rate can add 10 to 100 times more heat to the body than radiation and convection combined. In the case of cold stress, the metabolic rate affects heat balance to the same extent as radiation and convection loss.

- Clothing has three relevant characteristics affecting thermal balance: insulation, permeability, and ventilation.

Insulation

- Insulation is a measure of the resistance to heat flow by radiation, convection, and conduction. The greater the amount of insulation, the less the rate of heat flow from the warmer mass to the cooler one.

- During heat stress exposures, insulation reduces heat flow by radiation and convection.

- In cold stress, insulation reduces heat loss by convection and radiation as well as conduction. Insulation also prevents cold injury to tissues by preventing direct contact with cold surfaces.

Permeability

- Permeability measures the resistance to water vapor movement through the clothing. It is a factor in thermal stress because it influences the amount of evaporative cooling that takes place.
- Permeability is related to both insulation characteristics and clothing fabrics. As insulation increases, permeability decreases. Some contamination barriers reduce the magnitude of permeability. There is a trade-off between the risks of heat stress and the risks from skin contact with harmful chemicals. New personal protective clothing attempts to address these issues.

Ventilation

- Clothing ensembles that enable the movement of air enhance evaporative and convective cooling. Garments that are designed to limit air movement limit evaporative and convective cooling.

Heat Stress

HEAT STRESS PHYSIOLOGY

- Heat stress is caused by a combination of factors including climate, work demands, and clothing. Heat stress raises body temperature, heart rate, and increases sweating.

Terms Necessary for Heat Stress Analysis

- Heat stress: A combination of environmental factors, work demands, and clothing that constitute the total heat load imposed on the body, leading to increased body temperature, increased heart rate, and increased sweating.
- Heat Strain: Collective physiological adaptations and responses to heat stress.

Muscular Work

- Heat generated by muscular work heats the deep body tissues. This increases the body's core temperature.
- An increased heart rate reflects the demands on the cardiovascular system to move blood (and heat) from the body core to the skin.
- The greater the level of heat stress, the greater the sweat rate.

Heat Tolerance Factors

- Heat tolerance factors depend on many variables, including:
  - Climatic conditions.
  - Work demands.
  - Clothing.
  - Acclimation (usually two days to two weeks is required for full acclimatization).

- Skin area to weight ratio.
- Age and disease (with increased age, there is decreased onset of sweat).
- Water and salt balance.
- Alcohol consumption.
- Physical fitness.

Heat-Related Disorders

- Heat rash: Tiny, raised vesicles on affected areas.

- Heat Cramps: Painful spasms of voluntary muscles due to loss of salt in the muscles.

- Heat Fatigue: Impaired performance of mental tasks.

- Heat Exhaustion: Profuse sweating, rapid pulse, nausea, weakness, dizziness, cool and clammy skin, with normal oral temperature, but elevated core temperature.

- Heat Stroke: Although rare, most serious because risk of death is high. Characterized by diminished sweating, hot, dry, flushed skin, and increased temperature. Cool potential heat stroke victim immediately.

RECOGNITION OF HEAT STRESS

- Recognize heat stress by physiological markers and work behaviors. Other less recognizable signs include hypothalamic regulation and increased core temperature.

- Physiological markers include:
  - Elevated oral temperature.
  - Elevated heart rate.
  - Water loss.

- Worker behaviors that indicate heat stress include:
  - Adjusting/removing clothing.
  - Slowing work rate.
  - Taking small breaks.
  - Using shortcuts.
  - Irritability.

- Hypothalamic regulation (36-38°C) of body temperature includes:
  - Perspiration—cooling by evaporation.
  - Vasodilation—increased peripheral circulation.

- Core temperatures should not exceed 100.4° F (38° C).

Estimating Heat Stress

- Methods to evaluate heat stress require an assessment of metabolic rate and some measure of the thermal environment.

Assessment of Metabolic Rate

- The metabolic rate is the rate of internal heat generation.
- Assess metabolic rate by classifying the work demands into three (light, moderate, and heavy) to five (very light, light, moderate, heavy, and very heavy) categories of metabolic rate.
- Another method to assess metabolic rate is to look for similar activities in published tables of metabolic rates for specific activities. Both methods are expedient, but not accurate.
- A third method is to use the NIOSH/ACGIH method, dividing the job being assessed into discrete, homogeneous tasks, then determining the frequency and duration of each. Once a metabolic rate can be assigned to each task, the time weighted average for metabolic rate can be determined (TWA-M). Refer to the ACGIH TLV booklet for details.

Assessment of Environmental Conditions

- To assess environmental conditions, essential factors include the air temperature, humidity, air speed, and average temperature of solid surroundings.
- Instruments used to measure environmental conditions include dry-bulb thermometers, wet-bulb thermometers, globe temperatures, and air velocity and integrating devices.

Wet-bulb Thermometer

- The psychrometric wet bulb temperature (T_{pwb}) is based on the degree of evaporative cooling.
  - A wetted wick is wrapped around a temperature sensor and enough air (>3 m/s) is forced over the wick to maximize the rate of evaporative cooling.
  - Temperature reduction is dependent upon water vapor in the air.
- The natural wet bulb (T_{nwb}) is similar to psychometric wet bulb.
  - Air is allowed to flow over the sensor naturally rather than being forced.

Dry-bulb Thermometer

- Dry-bulb Thermometer (T_{db}) is a direct measure of air temperature.

Globe Temperature (Tg)

- The globe temperature is a measure of radiant heat from the solid surroundings and convective heat exchange with the ambient air. The globe temperature is used to estimate the average wall temperature of the surroundings.
  - It is measured using a six inch thin-walled copper sphere painted matte black on the outside.
  - A temperature sensor is placed in the center of the globe.
  - The globe is suspended in the air.

Air Velocity (Vair)

- Air velocity is a measure of the average measured air speed using an anemometer (not a unidirectional anemometer).

Integrating Devices

- Electronic devices measure wet bulb globe temperature (WBGT), integrating the two types of thermometers in one instrument.

Effective Temperature and Corrected Effective Temperature

- The effective temperature (ET) was first used as an index to quantify subjective thermal sensation. Determined from a nomogram, it was used in early studies of heat stress and did not account for radiant heat.

- Corrected effective temperature (CET) was developed to replace effective temperature, since ET did not account for radiant heat.

- ET and CET are no longer used to evaluate heat stress. Wet bulb globe temperature (WBGT) has replaced them.

Wet Bulb Globe Temperature (WBGT)

- The wet bulb globe temperature (WBGT) is an index of environmental heat, widely used either indoors or outdoors.

- Indoors or outdoors on a cloudy day, WBGT can be calculated by:

$$WBGT_{in} = 0.7\ T_{nwb} + 0.3\ T_g$$

- Outdoors under conditions of direct sunlight and no cloud cover, WBGT can be calculated as follows:

$$WBGT_{out} = 0.7\ T_{nwb} + 0.2\ T_g + 0.1\ T_{db}$$

where:

WBGT = indoor or outdoor wet-bulb globe temperature
T_{nwb} = natural wet bulb temperature (evaporation)
T_{db} = dry bulb temperature (air temperature—convection)
T_g = globe temperature (radiation)

Exposure Standards

- Exposure standards for heat stress are outlined by ACGIH.

- The ACGIH TLV establishes that deep body temperature should not exceed 38°C (100°F). WBGT, workload, and a work/rest regimen are used to determine the time-weighted-average (TWA).

$$\text{TWA TLV} = \frac{(\text{WBGT}_1 \times t_1) + (\text{WBGT}_2 \times t_2) + \ldots (\text{WBGT}_n \times t_n)}{t_1 + t_2 + t_3 \ldots + t_n}$$

- These values are for workers who are wearing standard long-sleeved, long-trousered, cotton/synthetic work clothes.

- An increase in the insulation value and/or decrease in vapor permeability will reduce the amount of heat that can be lost. For such workers, it is necessary to monitor heart rate and body temperature. Intervals are dependent on level of physical work and adjusted temperature. (*The Occupational Health & Safety Guidance Manual for Superfund Activities* published by the US Department of Health and Human Services provides a Time vs. Adjusted Temperature nomogram for semi-permeable and impermeable clothing).

HEAT BALANCE ANALYSIS

- Heat balance analysis uses a model of the exchange between a hypothetical person and the environment. Whereas WBGT methods previously described are empirical methods of heat stress evaluation, heat balance analysis is considered a rational method.

- Heat balance analysis helps determine the amount of time it takes to reach an upper limit of thermal heat storage (38 °C) and evaluations of countermeasures can be performed. Heat balance analysis can be performed using the heat stress index.

Heat Stress Index

- Heat Stress Index (HSI), also known as the Belding-Hatch heat stress index, combines the environmental heat and metabolic heat into an expression of stress in terms of requirement for evaporation of sweat.

$$\text{HSI} = 100 \times \frac{E_{req}}{E_{max}}$$

where:

HSI = heat stress index expressed in %
E_{req} = (M + R + C) required rate of heat loss by evaporation
E_{max} = maximum evaporative heat loss or maximum evaporative capacity of the environment (BTUs/hr or W/m^2)

Radiant Heat Transfer

- Radiant heat transfer is frequently caused by infrared radiation being emitted from a heated source.

- Determining radiant heat transfer requires several steps.
  - Calculate the mean radiant (wall) temperature and then insert the information into the radiant heat formula.
  - A negative number symbolizes heat loss by a worker and a positive number signifies a heat gain or possible heat stress situation.
  - The following equation is used:

$$R = 17.5(t_w - 95)$$

where:

$$t_w = t_g + 0.13v^{0.5}(t_g - t_a)$$

and:

R = radiant heat transfer in BTU/hour (substitute 15 if workers are fully clothed and use 25 for seminude workers wearing shorts without shirts).
t_w = mean wall temperature in degrees F
v = air velocity in fpm
t_g = globe temperature in degrees F
t_a = air temperature in degrees F

- Note: 95 refers to the mean skin temperature in degrees F.

Convective Heat Transfer

- Convective heat transfer is based on the movement of fluid around a source of heat and consequently transferring part of that heat to another object.

- It is expressed as:

$$C = 0.756v^{0.6}(t_a - 95)$$

where:

C = convective heat transfer in BTU/hr
v = air velocity in fpm
t_a = air temperature in degrees F

- Note: Use the 0.756 factor when no other factor is provided. Use 0.65 for clothed workers (shirt and trousers) and 1.08 for semi-nude workers (shorts but no shirts).

Maximum Evaporative Capacity

- The maximum evaporative capacity (E_{max}) measures the cooling capacity of the average human body and offsets metabolic and radiant or convective heat loads.
- It is the maximum amount of BTUs that can be eliminated by the worker per hour under the given conditions of air speed, temperature, and moisture content in air.
- If the workplace atmosphere is completely saturated at its given temperature, then no evaporative capacity is possible.
- The maximum evaporative capacity is expressed as:

$$E_{max} = 2.8\, v^{0.6}(42 - WP_w)$$

where:

E_{max} = maximum evaporative capacity (BTU/hr)
v = air velocity in feet per minute
VP_w = water vapor pressure in mm Hg

Note: Use 2.8 when no specific information is available. Use 4.0 for semi-nude workers with shorts but no shirt, and 2.4 for fully clothed workers with trousers and shirt.

GENERAL CONTROL MEASURES

- General control measures for heat stress include:
  - Training.
  - Fluid replacement. (Thirst is not a sufficient driver for fluid replacement. Drinking small quantities of water frequently is recommended.)
  - Self-determination.
  - Diet.
  - Lifestyle.
  - Health status.
  - Acclimation (acclimatization).

Engineering Controls for Heat Stress

- Engineering control measures for heat stresses include a variety of methods.
- Reduce physical work demands, using power tools whenever possible.

- Control air temperature. When air temperature is above 104° F or 40° C, workers gain heat from the air. When air temperature is below 90° F (32° C), there is significant loss of body heat.
  - Dilution ventilation brings in a supply of cool air.
  - Active cooling, evaporative cooling, and mechanical refrigeration can be used in the work area.

- Reduce air humidity by removing water vapor from the air with mechanical refrigeration in the area.

- Change clothing.

- Reduce radiant heat. When globe temperature is greater than 109° F (43° C), radiant heat is a significant source of heat stress.
  - Insulate surfaces to reduce surface temperatures.
  - Provide shielding, if source is well defined.

- Increase air movement.
  - This enhances evaporative cooling and convective cooling if the air temperature is less than 95° F (35° C).
  - Between 95° and 104° F (35° and 40° C), heat gain by convection increases with increases in air movement.
  - Above 104° F (40° C), increases in air movement actually increases overall heat stress.
  - The chief method for increasing air movement is to use a fan in the workplace (with a recommended rotation of two meters per second).

ADMINISTRATIVE CONTROLS FOR HEAT STRESS

- Administrative controls for heat stress include:
  - Acclimatization.
  - Pacing of work.
  - Sharing of work.
  - Scheduling of work responsibilities.

Personal Protection

- Personal protection methods for control of heat stress include:
  - Circulating air systems.
  - Circulating water systems.
  - Ice garments.
  - Reflective clothing.

Cold Stress

COLD STRESS FACTORS

- Cold stress is caused by climatic factors, such as air temperature, wind speed, and wind chill. Cold stress factors account for achieving an effective chill temperature (ECT).
- Other cold stress factors include age, disease, and overall physical condition.

Cold Stress Physiology

- The physiology of cold stress includes vasoconstriction, reduced peripheral circulation, and increased flow to vital organs. This is accompanied by pain in extremities, which is usually the first warning of cold stress.
- There is a relationship between clinical signs and core temperature:

Clinical Sign	Core Temperature (°C)
Normal oral breathing	37
Maximum shivering	35
Hypothermia begins	33
Consciousness fades, weak pulse, shivering ceases, pupils dilated	32
Loss of consciousness	26
Heart Stops	20

Exposure Standards for Cold Stress

- The ACGIH TLV considers work load, work warm-up regime, temperature, and wind speed for cold stress exposure standards.
- The objective is to keep the deep body temperature from falling below 36° C and to prevent injury to the extremities.

EVALUATION OF COLD STRESS

- ACGIH recommends protective clothing for temperatures less than 41° F (5° C). To estimate the amount of clothing insulation required (in clo units), the following formula is used:

$$I_{clo} = \frac{11.5\,(33 - T_{db})}{M}$$

where:

I_{clo} = clo units (1 clo = 0.156 m^2 x C/W)
T_{db} = dry bulb temperature (° C)
M = metabolic rate in watts

CONTROL MEASURES FOR COLD STRESS

- Control measures for cold stress include:
  - Heating (using warming devices, shelters, hand warmers).
  - Minimizing air velocity.
  - Adjusting length of exposure.
  - Insulating workers (type of clothing).
  - Insulating convective sources (e.g., metal handles).
  - Supplying warm drinks, such as soups (but not coffee).
  - Protective observation (e.g., the buddy system).
  - Modifying work schedules.
  - Education.
  - Medical surveillance.

12 Noise and Noise Control

High noise levels in the workplace are a physical stress that may produce psychological effects by annoying, startling, or disrupting the concentration of a worker. High levels can also damage workers' hearing. Industrial hygienists must be familiar with noise, its sources, and its adverse health effects on workers to create effective control measures.

Basic Concepts

NOISE TERMINOLOGY

- A variety of specialized terms are used to express concepts in noise and noise control.

Noise and Sound

- Noise and sound are usually used interchangeably, but sound is descriptive of useful communication or pleasant sounds, while noise is used to describe discord or unwanted sound.

Sound Intensity (I)

- Sound intensity at a specific location is the average rate at which sound energy is transmitted through a unit area normal to the direction of sound propagation.
- Units are joules per square meter per second. Units may also be expressed in terms of a level (i.e., sound intensity level L_I) in decibels referenced to 10^{-12} watts per square meter.

Sound Power (w)

- Sound power of a source is the total sound energy radiated by the source per unit time.
- It is expressed in terms of the sound power level (L_w) in decibels referenced to 10^{-12} watts (w_0).
- The relationship to decibels is shown below:

$$L_w = 10 \log \frac{w}{w_0} \quad \textbf{or} \quad L_w = 10 \log w + 120$$

where:

L_w = sound power level (decibels)
w = sound power (watts)
w_0 = reference power (10^{-12} watts)
log = a logarithm to the base 10

- Sound power cannot be measured directly. It is possible to measure intensity, but the instruments are very expensive.

Sound Pressure (SPL)

- Sound pressure normally refers to the root mean square (rms) value of the pressure changes above and below atmospheric pressure when used to measure steady-state noise.

- Units used to describe sound pressures are newtons per square meter (N/m^2), dynes per square centimeter ($dyne/cm^2$), microbars (μbars), and micropascals (μPa).

  $$1\ \mu bar = 1 dyne/cm^2 = 0.1\ N/cm^2 = 0.1\ Pa$$

- Sound pressure is also described in terms of a level (sound pressure level, SPL) in decibels referenced to 0.00002 N/m^2 or 0.0002 $dynes/cm^2$ or 0.0002 μbars.

- Sound pressure can be measured more easily than sound power levels, so sound level meters measure sound pressure levels (SPL) in decibels.

- The sound pressure relationship to decibels is:

$$SPL = 10 \log \frac{p^2}{p_0^2} \quad \textbf{or} \quad 20 \log \frac{p}{p_0}$$

where:

SPL = sound pressure level (decibels)
p = measured root-mean-square (rms) sound pressure (N/m^2, μbars)
p_0 = reference rms sound pressure (20 μPa, N/m^2, μbars)

- Shortcuts: dB = 20 log p + 94 for N/m^2
  dB = 20 log p + 74 for microbars

Velocity (c)

- The speed at which the regions of sound producing pressure changes move away from the sound source is called the velocity of propagation.

- Sound velocity varies directly with the square root of the density and inversely with the compressibility of the transmitting medium, as well as with other factors.

- Velocity of sound is constant in a given medium over the normal range of conditions.

- Velocity of sound is approximately 1126 ft/sec in air, 331 m/sec at STP, (1130 ft/sec in air, or 344 m/sec at NTP) 4700 ft/sec in water, 13,000 ft/sec in wood, and 16,500 ft/sec in steel.

- The speed of sound is always equal to the product of the wavelength and the frequency:

  $c = f\lambda$

Wavelength (λ)

- The distance required for one complete pressure cycle to be completed is called one wavelength and is measured in feet or meters.

- The wavelength (λ) may be calculated from known values of frequency (f) and velocity (c):

  $\lambda = c/f$

- Sound waves that have a wavelength much larger than an obstacle are little affected by the presence of that obstacle; the sound waves bend around it. Bending of sound waves is called diffraction.

- If the wavelength of sound is small in comparison with the size of an obstacle, the sound is reflected or scattered in many directions around that object. As a result, a wall is of little use as a shield against low frequency sound, but it can be an effective barrier against high frequency sound.

Directivity Factor (Q)

- Directivity (Q) is a measure of the degree to which sound is concentrated in a certain direction, rather than radiated evenly in a full spherical pattern.

- Directivity factors can be derived from tables. They are actually portions of spherical radiation patterns as related to the surface area of a sphere ($4\pi r^2$).

Free field

- A free field exists in a medium free from boundaries.

- In a free field, sound radiated from a source can be measured accurately without influence from the test space because of reflecting surfaces.

- True free fields are rarely found, except in echo chambers.

- Sound pressure decreases inversely in a free field.

- For free field radiation, directivity equals one (Q = 1).

Period (T)

- The period is the time required for one cycle of pressure change to take place. It is reciprocal to frequency, and is measured in seconds.

Frequency

- Frequency describes the rate at which complete cycles of high and low pressure regions are produced by the sound source.
- Units are expressed as hertz (Hz) and formerly, cycles per second (cps).
- The audible range of frequencies for humans with good hearing is between 20 to 20,000 cps at moderate sound levels. As humans age, they lose their sensitivity to higher frequencies (presbycusis).
- Frequency will increase or decrease as the distance from the source decreases or increases. This is called the Doppler effect.
- A sound's frequency composition is termed its spectrum which can be a determining factor in the level of annoyance caused by noise. High frequency noise is generally more annoying than low frequency noise.

Pitch

- Pitch is a measure of auditory sensation that depends primarily upon frequency.
- Example: In a hearing booth, you hear a 2000 Hz tone, then a 4000 Hz tone. You perceive that the pitch has gone up. The ability to hear is a sensation in the brain in response to sound waves.

Reverberation

- Reverberation occurs when sound persists after the direct reception of the sound has stopped.
- Reverberation time is the time required after the source has stopped radiating sound for the root mean square sound pressure to decrease 60 dB from its steady-state level.

Root-Mean-Square (rms) Sound Pressure

- The root-mean-square (rms) value of a changing quantity, such as sound pressure, is the square root of the mean of the squares of the instantaneous values of the quantity. This value is used in electrical theory as well.

Noise Measurement

MEASURING AND CALCULATING NOISE EXPOSURE

- A number of instrument measurements are used to determine occupational exposure to noise, including sound power and sound pressure. The unit of measurement for sound is the decibel.

Decibels (dB)

- Decibels are the preferred unit for measuring sound, derived from the bel, a unit of measure in electrical communications engineering.
- The decibel is a dimensionless unit used to express the logarithm of the ratio of a measured quantity to a reference quantity.
- In acoustics, the decibel is used to describe the level of quantities that are proportional to sound power.
- Decibel is not only used in the description of sound pressure level and sound power level, it can also be used to describe hearing thresholds, electrical voltage, electrical current, acoustic intensity, and electrical power.

SOUND LEVEL METERS

- The basic instrument used to measure sound pressure variations in air is the sound level meter (SLM).
- The SLM contains a microphone, an amplifier with a calibrated attenuator, a set of frequency-response networks (weighting networks), and an indicating meter.
- Sound level meters measure the electrical signal emitted from a microphone. The signal is amplified sufficiently and converted to direct current to deflect a needle on an indicating meter. The measurement range is 40-140 dB.
- Designed for field use, the SLM is reliable, rugged, reasonably stable under battery operation, and lightweight.

OCTAVE BAND ANALYZERS (OBA)

- Use an octave band analyzer (OBA) to determine where the noise energy lies in the frequency spectrum.
- An OBA breaks the total noise down into its various frequency components—low, middle, and high frequency.
- OBAs are necessary because:
  - People react differently to low frequency and high frequency noise at the same sound pressure level.
  - The engineering solutions for reducing or controlling noise are different for low frequency and high frequency noise (low frequency is more difficult to control).

- Frequency distribution is determined by passing noise successively through several different filters that separate the noise into eight or nine octaves on a frequency scale.

- Most OBA filters cover the audio range from about 22 Hz to about 11,300 Hz in nine octave-frequency bands. Narrower bands are used for more detailed analysis and divide the band into three parts.

Commonly Used Octave Bands

Center Frequency of Band (Hz)	Frequency Range of Band (Hz)
31	22 - 44
63	44 - 88
125	88 - 175
250	175 - 350
500	350 - 700
1000	700 - 1400
2000	1400 - 2800
4000	2800 - 5600
8000	5600 - 11200

Calculation of Frequencies

- Calculation of frequencies can be performed using several formulae, including calculations for the upper frequency band, one-half octave band, and one-third octave band.

Upper Frequency Band

- The upper frequency band can be calculated as:

$$f_2 = 2f_1$$

where:

f_2 = upper frequency band
f_1 = lower frequency band

One-Half Octave Band

- The one-half octave band can be calculated as follows:

$$f_2 = \sqrt{2(f_1)}$$

where:

f_2 = 1/2 octave band
f_1 = lower frequency band

One-Third Octave Band

- The one-third octave band can be calculated as:

$$f_2 = \sqrt[3]{2}(f_1)$$

where:

f_2 = 1/3 octave band
f_1 = lower frequency band

NOISE DOSIMETERS

- Noise dosimeters are not like sound level meters that measure discrete noise levels. Noise dosimeters integrate all noise levels into a time-weighted-average (TWA).

- Use noise dosimeters when workers move about several locations in the course of their duties or perform a variety of operations during the day and are subjected to different noise levels.

- Place the microphone of the dosimeter in the worker's hearing zone. Place the dosimeter on the person's belt. The dosimeter computes the noise measurements and daily noise dose that the worker would experience.

- Compare these results to the OSHA standard.

ADDING NOISE SOURCES TOGETHER

- Industrial hygienists often must evaluate workplace noise and add two independent noise sources together for an overall sound pressure level.

- Because decibels are logarithmic values, it is not possible to add them by normal algebraic addition. 60 dB plus 60 dB does not equal 120 dB, but rather 63 dB.

Sound Power Additions

- When sound power is known, the formula for adding noise sources is:

$$L_W = 10 \log \frac{(w_1 + w_2)}{(w_0 + w_0)}$$

where:

L_w = sound power in watts
W_1 = sound power of noise source 1 in watts
w_2 = sound power of noise source 2 in watts
w_0 = reference sound power (reference 10^{-12}) watts

Sound Pressure Additions

- When sound pressure is known, the formula for adding noise sources is:

$$SPL = 10 \log \frac{p^2}{p_0^2}$$

where:

$$\frac{p^2}{p_0^2} = 10^{SPL/10}$$

and:

SPL = sound pressure level (decibels)
p = measured root-mean-square (rms) sound pressure (N/m², μbars)
p_0 = reference rms sound pressure (20 μPa, N/m², μbars)

- For three sources, the equation becomes:

$$SPL = 10 \log (10^{SPL_1/10} + 10^{SPL_2/10} + 10^{SPL_n/10})$$

- When adding any number of sources the equation becomes:

$$SPL = 10 \log (10^{SPL_1/10} \;.....+\; 10^{SPL_n/10})$$

- This formula can be used for numerous noise sources, whether the sources are identical or not.

Adding Identical Sources

- Use this simplified formula for determining the sound pressure level from multiple identical sources.
- This assumes that each source produces the same SPL at the listener's ear.

$$SPL_f = SPL_i + 10 \log n$$

where:

SPL_f = total sound pressure level (dB)
SPL_i = individual sound pressure level (dB)
n = number of identical sources

Calculation of Noise Levels in a Free Field

❑ The calculation for noise levels in a free field is expressed as:

$$SPL = L_w - 20 \log r - 0.5$$

where:

SPL = sound pressure (reference 0.00002 N/m2)
L_w = sound power (reference 10^{-12} watts)
r = distance in feet

Calculation of Noise Levels with Directivity Factors

❑ Calculations for noise levels with directional characteristics is expressed as:

$$SPL = L_w - 20 \log r - 0.5 + 10 \log Q$$

where:

SPL = sound pressure (reference 0.00002 N/m²)
L_w = sound power (reference 10^{-12} watts)
r = distance in feet
Q = directivity factor
Q = 2 for one reflecting plane
Q = 4 for two reflecting planes
Q = 8 for three reflecting planes

Sound Pressure Level as a Function of Distance from a Point Source

❑ Calculating the noise level at a new distance from the noise source can be computed as follows:

$$SPL = SPL_1 + 20 \log \frac{(d_1)}{(d_2)} \quad \textbf{or} \quad SPL = SPL_1 - 10 \log \frac{(d_2)^2}{(d_1)^2}$$

where:

SPL = sound pressure level at new distance (d_2)
SPL_1 = sound pressure level at d_1
d_n = distance from source

Adding Decibels from a Table

❑ The following table represents a simplified schedule for adding decibels:

Difference in decibel values	Add to higher value
0 or 1 dB	3 dB
2 or 3 dB	2 dB
4 to 9 dB	1 dB
10 dB or more	0 dB

WEIGHTING NETWORKS

- Three different internationally standardized characteristics (called weighting networks A, B, and C) were developed that vary with frequency in the same way as the human ear.
  - The A-network was designed to approximate the equal-loudness curves at low sound pressure levels.
  - The B-network was designed for medium sound pressure levels.
  - The C-network was designed for high sound pressure levels.

- The weighting networks are the sound level meter's means of responding to some frequencies more than to others.

- Low frequencies are discriminated against severely by the A-network, moderately attenuated by the B-network, and barely attenuated by the C-network. All three networks respond similarly at the 1000 Hz frequency. Example: If the measured sound level of a noise is much higher on C-weighting than the A-weighting, much of the noise is probably low frequency.

- By definition, a weighted-frequency scale is simply a series of correction factors that are applied to sound pressure levels on an energy basis as a function of frequency.

The A Scale

- The A-weighted sound level measurement is thought to provide a rating of industrial broadband noises that indicates the injurious effects such noise has on the human ear. Measurements are expressed in dBA.

- ACGIH and OSHA have adopted the A-weighted sound levels as the measurement for assessing noise exposure.

- This scale has been shown to provide reasonably good assessments of speech interference and community disturbance conditions and has been adopted by the EPA.

Sound Surveys and Noise Standards

TYPES OF SOUND SURVEYS

- Two broad categories of sound surveys are source measurements and ambient-noise measurements.

- Use source measurements to collect acoustic data for determining the characteristics of noise radiated by a source and describing the characteristics of a particular noise source.

- Use ambient-noise measurements to study a single sound level or to make a detailed analysis.

- Ambient-noise measurements describe the characteristics of a sound field of largely unspecified or unknown sources.

Conducting Sound Surveys

- Conduct preliminary noise surveys in the workplace to identify areas that need detailed study.
  - Conduct detailed noise surveys.
  - Use a general-purpose sound level meter for preliminary studies and dosimeters or OBAs for more detailed studies.
  - Set the sound level meter to A-scale slow response for OSHA compliance purposes.
  - Take measurements at the ear level.

The OSHA Noise Standard

- The OSHA noise standard (29 CFR 1910.95) establishes permissible employee noise exposures in terms of duration (in hours per day) at various sound levels.

- Employers must reduce exposures through engineering or administrative controls.

- The standard defines the permissible exposure limit (PEL) as that noise dose that would result from continuous eight-hour exposure to a sound level of 90 dBA. This is a dose of 100 percent.

- Doses for other exposures that are either continuous or fluctuating in level are computed relative to the PEL based on a 5 dBA trading relationship between noise level and exposure time.

- Every 5 dBA increase in noise level cuts the allowable exposure time in half. This is known as the 5 dB exchange rate.

Continuous Noise

- Continuous noise is defined as broadband noise of approximately constant level and spectrum to which an employee is exposed for a period of eight hours per day, 40 hours per week.

Intermittent Noise

- Intermittent noise is defined as exposure to a given broadband sound-pressure level several times during a normal working day.

Impact Noise

- Impact noise is defined as a sharp burst of sound. Sophisticated instrumentation is necessary to determine the peak levels for this type of noise. This includes noise from hammer blows, explosions, punch presses, etc.
- Employees should not be exposed to impact noise that exceeds 140 dB.
- ACGIH recommends that impact noise be permitted as follows:

Sound level (dB)	Permitted number of impacts per day
140	100
130	1000
120	10000

The Ears, Hearing Loss, and Calculating Exposures

THE EAR

- The ear consists of three parts: the external ear, the middle ear, and the inner ear.
- The external ear consists of two sections: the auricle (or pinna) and the external auditory canal.
- The middle ear is the cavity between the eardrum and the bony wall of the inner ear. It consists of the Eustachian tube, the eardrum, ossicular chain, oval and round windows, and the mastoid air cell system.
- The inner ear contains receptors for hearing and position sense, and the cochlea, vestibule, and semicircular canals.

The Hearing Process

- The hearing process occurs when the outer ear funnels and conducts sound vibrations to the eardrum through the canal.
  - The eardrum vibrates in response to the sound waves that strike it.
  - This vibratory process, in turn, is transmitted to the chain of three tiny bones in the middle ear.
  - The ossicles conduct the sound vibration across the air-filled middle ear cavity to a fluid in the delicate inner ear.
  - The vibration of the ossicles create waves in the inner ear fluid that stimulate microscopic hair cells.
  - The stimulation of these hair cells generates nerve impulses, which pass along the auditory nerve to the brain for interpretation.

HEARING LOSS

- The outer and middle sections of the ear conduct sound energy to the deeper structures. Therefore, the outer and middle area together act as the conductive hearing mechanism.
- The deeper structures, including the inner ear and the auditory nerve, are referred to as the sensorineural mechanism.
- Conductive and sensorineural describe two major types of hearing loss.

Conductive Hearing Loss

- When test results show that a person has depressed hearing by air conduction but normal hearing by bone conduction, the presence of conductive hearing loss is indicated.
- A conductive loss is due to some impairment of sound transmission before it reaches the inner ear. It results from interference with the function of the outer or middle ear.
- Causes of conductive hearing loss include ear wax build-up and infection.

Sensorineural Hearing Loss

- If a person hears just as poorly by bone conduction as by air conduction, then the hearing loss can be due to damage in the deep structures of the ear.
- Hearing impairment that involves only the inner ear or the auditory nerve is classified as sensorineural impairment. This can involve impairment of the cochlea, the auditory nerve, or both.
- Causes of sensorineural hearing loss include certain antibiotics (aminoglycosides), rare but permanent complication of mumps, damage to ear structures from blows and skull fractures damaging ear structures, and long term damage to the hair cells of the cochlea from noise.

Noise-Induced Hearing Loss

- Permanent damage from noise is generally classified as noise-induced hearing loss.
- When an individual is exposed to high noise levels throughout a workday, hearing loss occurs gradually throughout the day. This phenomenon is called a temporary threshold shift. After a rest period away from the noise, the hearing usually returns to its former level. A rest period of 14 hours or so away from the noise is adequate to return the threshold to previous levels.

- Long-term cumulative effects of repeated and prolonged hazardous noise exposure result in permanent pathological changes in the cochlea and irreversible threshold shifts in hearing acuity.

- An initial change in hearing is usually observed in the higher frequency range, at about 4000 Hz. Noise-induced hearing loss is usually represented by a notch at the 4000 Hz frequency on the audiogram (with O used to denote the right ear, X for the left ear). Further exposure deepens and widens the notch. When hearing loss involves the speech frequency range, considerable difficulty in hearing conversational speech results.

Effects of Noise Exposure

- Noise induced hearing loss involves damage to the cochlea.

- Temporary Threshold Shift (TTS) is a temporary hearing loss.
  - TTS can be produced by brief exposures to high level sound.
  - Sound levels must exceed 60-80 dB before a person develops TTS.
  - The longer the sound lasts, the greater the amount of threshold shift.
  - TTS is greatest immediately after exposure to excessive noise and progressively diminishes with increasing rest time as the ear recovers from the apparent noise overstimulation.
  - Sound having most of its energy in the speech frequencies is more potent in causing TTS than low frequencies.
  - The longer and more numerous the quiet periods between periods of noise, the lower the potential for threshold shifts.

Analysis of Noise Exposure

- Any exposure of the unprotected ear to sounds > 115 dBA is hazardous. Exposures less than 70 dBA can be assumed safe.

- Noise energy with predominant frequencies > 500 Hz has a greater potential for causing hearing loss than lower frequencies.

- Pure tone frequencies are believed to be more hazardous to hearing than equal energies across a broad band frequency range.

- Incidence of noise-induced hearing loss is directly related to total exposure time.

- Intermittent exposures are far less damaging to the ear than are continuous exposures, even if the sound pressure levels are higher.

RISK FACTORS

- Factors that affect the degree and extent of hearing loss, include:
  - Intensity of the noise (sound pressure level).
  - Type of the noise (frequency spectrum).
  - Period of exposure.
  - Total work duration (years of employment).
  - Individual susceptibility.
  - Age of the worker.
  - Coexisting hearing loss and ear disease.
  - Character of the surroundings in which the noise is produced.
  - Distance from the source.
  - Position of the ear with respect to sound waves.
- Twenty percent of the general population between the ages of 50 and 59 experience hearing losses without having had any exposure to industrial noise, but groups of workers exposed to steady-state industrial noise over a working lifetime show a more pronounced incidence of hearing loss.

TESTING OF HEARING ABILITY

- Two classes of tests that evaluate the ability to hear speech include:
  - Those that measure the hearing threshold or the ability to hear very faint speech sounds.
  - Those that measure discrimination, or the ability to understand speech.
- The average hearing level for pure tones at 500 Hz, 1000 Hz, and 2000 Hz is used as an indirect measure of the probable ability to hear everyday speech.
- If the average monaural hearing level at 500, 1000, and 2000 Hz is 25 dB or less, the American Medical Association (AMA) has suggested that no impairment in the ability to hear everyday speech occurs under everyday conditions.

Audiometry

- Audiometry, the measurement of hearing, is a vital part of a comprehensive hearing conservation program. Selection for audiometry generally includes employees in a hearing conservation program and pre-placement employees.
- Use audiometry to:
  - Obtain a baseline audiogram that indicates an individual's hearing ability at the time of the preplacement examination.
  - Provide a record of an employee's hearing acuity. Check the effectiveness of noise-control measures by measuring the hearing thresholds of exposed employees.

- Record significant hearing threshold shifts in exposed employees during the course of their employment.
- Comply with government regulations.

Suitable Test Environments

- Use an audiometer to assess an individual's hearing ability. This instrument should meet the standards set forth by ANSI S3.6-1989, Specifications for Audiometers.
- Take hearing measurements in a room or booth that conforms to the requirements established by ANSI S3.1-1991, Maximum Permissible Ambient Noise Levels for Audiometric Test Rooms. It must be sufficiently quiet within the enclosure so that external noises do not interfere with the employee's perception of the test sounds.
- Locate booths away from outside walls, elevators, and locations with heating and plumbing noises.

Adequate Records

- Use medical forms in audiometric testing programs that include all basic data related to the hearing evaluation, hearing threshold values, noise-exposure history, and pertinent medical history.
- Keep records for at least the duration of employment.

Noise Exposure Regulations

OCCUPATIONAL NOISE STANDARD

- The OSHA standard, 29 CFR 1910.95, requires employers to reduce exposure to 90 dBA over an eight-hour workday through engineering or administrative controls.
- In all cases where the sound level exceeds the PEL, regardless of the use of hearing protective devices, "a continuing and effective hearing conservation program" is required.

Hearing Conservation Program

- An effective hearing conservation program prevents hearing loss as a result of occupational noise exposure.
- It limits the amount of compensable hearing loss in the frequency range over which normal hearing is necessary for communication. Compensable loss at present does not include frequencies over 4,000 Hz, although such loss impairs enjoyment of sound and may interfere with speech discrimination.

- A hearing conservation program must be instituted if any employee's noise exposure exceeds current limits as defined in the OSHA Noise Exposure standard.

- Although the eight-hour time TWA PEL remains at 90 dBA, a hearing conservation program becomes mandatory at an eight-hour TWA exposure of 85 dBA.

- All employees whose noise exposures equal or exceed 85 dBA must be included in a hearing conservation program comprised of five basic components:
  - Exposure monitoring.
  - Audiometric testing.
  - Hearing protection.
  - Employee training.
  - Recordkeeping (two years for surveys, duration of employment for audiograms).

Audiometric Evaluation

- Annual audiograms must be routinely compared to baseline audiograms to determine whether the audiogram is accurate and whether the employee has experienced lost hearing ability—that is, to determine whether a standard threshold shift (STS) has occurred.

- STS is defined as an average shift (loss) in either ear of 10 dB or more at the 2000, 3000, and 4000 Hz frequencies. A method of determining an STS by computing an average was chosen because it diminishes the number of persons identified as having as STS who are later shown not to have had a significant change in hearing ability.

Employees Identified with an STS

- Employees identified with an STS must be fitted or refitted with adequate hearing protectors. They must be notified within 21 days from the time the determination has been made.

- Some employees identified with an STS should be referred for further testing to distinguish hearing loss from a medical condition.

Calculating Daily Noise Dose (DND)

- Since sound pressure levels normally vary during a workshift, it is necessary to use a formula for compliance purposes. This formula combines the effects of different sound pressure levels and allowable exposure times.

- If the value exceeds 1, the standard has been exceeded.

$$DND = \frac{C_1}{T_1} + \frac{C_2}{T_2} + \frac{C_3}{T_3} ..... + \frac{C_n}{T_n}$$

where:

DND = Daily Noise Dose
C_i = number of hours exposed at given SPL_i
T_i = number of hours exposed at given SPL_i

Exposure (hr)	OSHA PELs (dB)
8	90
4	95
2	100
1	105
0.5	110
0.25	115

Calculating OSHA Permissible Noise Levels

- The formula below is used to determine the OSHA time limit for various sound pressure levels. Most people memorize the allowable time for given levels.

$$T_{SPL} = \frac{8}{2^{(SPL - 90)/5)}}$$

where:

T_{SPL} = time in hours at given SPL
SPL = sound pressure level (dBA)

TWA Equivalent from Noise Dose Measurements

- Noise dosimeters integrate sound pressure levels over time and produce readouts in terms of percentages of the eight hour PEL (i.e., 90 dBA).

- A value of 95 indicates that the exposure was at 95% of the OSHA allowable limit.

- Use the formula below to convert this value to the equivalent eight-hour TWA.

$$TWA_{eq} = 90 + 16.61 \log \frac{(D)}{(100)}$$

where:

TWA_{eq} = eight-hour equivalent TWA in dBA

D = noise dosimeter reading in %

Sample Problems

Sample Problem 1

❑ Determine the decibel level for a sound power of 0.001 watts.

$$L_w = 10 \log \frac{0.001}{10^{-12}}$$

$$L_w = 10 \log 10^9$$

$$L_w = 10 \times 9$$

$$L_w = 90 \text{ dB (re } 10^{-12})$$

Sample Problem 2

❑ What is the sound pressure level coming from a source which generates 1.57 N/m^2?

$$L_p = 20 \log \frac{1.57}{0.00002}$$

$$L_p = 20 \log 78{,}500$$

$$L_p = 20 \times 4.895$$

$$L_p = 97.9 \text{ dB (re 0.00002 N/m2)}$$

Sample Problem 3

❑ At a distance of 40 feet from a point source of noise, the sound level was measured at 90 dB. What is the level at 80 feet?

$$dB = 90 \text{ dB} - 20 \log \frac{80}{40}$$

$$dB = 90 \text{ dB} - 6$$

$$dB = 84 \text{ dB}$$

Sample Problem 4

- Two machines are each producing a 90 dB sound level. Calculate the combined sound level.

$$dB = 10 \log (10^{90/10} + 10^{90/10})$$

$$dB = 10 \log (2 \times 10^{9})$$

$$dB = 10 \times (0.3 + 9)$$

$$dB = 93$$

Sample Problem 5

- An employee was tested for eight hours and found subject to the following exposures: four hours at 80 dB, two hours at 85 dB, two hours at 90 dB. Calculate the noise exposure for the eight hours to determine exposure levels consistent with the relevant OSHA standards.

$$\frac{4}{32} + \frac{2}{16} + \frac{2}{8} = 0.5$$

No overexposure: 0.5 equals a daily noise dose (DND) of 50%.

Sample Problem 6

- What would be the sound pressure level produced in a free field at a distance of 75 feet directly in front of a machine? The directivity factor is 7. The noise source is continuous spectrum with a sound power of 0.2 watts.

$$L_p = 10 \log \frac{0.2}{10^{-12}} - 20 \log 75 - 0.5 + 10 \log 7$$

$$L_p = 10(\log 0.2 - \log 10^{-12}) - 20(1.9) - 0.5 + 10(.85)$$

$$L_p = 10(-0.7 + 12) - 38 - 0.5 + 8.5$$

$$L_p = 113 - 38 - 0.5 + 8.5$$

$$L_p = 83 \text{ dB}$$

Noise Control Programs

TYPES OF CONTROLS

- To control noise, understanding the source, path, and receiver are important to selecting and implement the correct control method.

Source

- The most desirable method of control is to minimize noise at the source. This can be achieved by modifying existing equipment and structures or possibly introducing noise-reduction at the design stage.

Path

- Modification along the noise path may be necessary when minimization at the source cannot be achieved or has been maximized.

- Modification along the noise path can be accomplished by:
  - Shielding or enclosing the source.
  - Increasing the distance between the source and the receiver.
  - Placing a shield between the source and the receiver.
  - Using baffles and enclosures placed over noise-producing equipment to minimize transmission of noise to occupied areas.
  - Using acoustical material on walls, ceilings, and floors to absorb sound waves and to reduce reverberations.

Control Measures

- Engineering controls include such solutions as:
  - Proper and preventive maintenance.
  - Substitution of machines.
  - Substitution of processes.
  - Reduction of the driving force of vibrating surfaces.
  - Reduction of the response of vibrating surfaces.
  - Reduction of the sound radiation from vibrating surfaces.
  - Reduction of the sound transmission through solids.
  - Reduction of the sound produced by gas flow.
  - Reduction of noise by reducing its transmission through air.
  - Isolation of the operator by means of a relatively soundproof booth.

- Administrative controls may be necessary when other methods fail. These include production schedule rotation.

- The OSHA standard specifies that engineering and administrative controls are to be used when feasible. If not feasible or pending, hearing protection is appropriate.

Enclosures

- When utilizing engineering controls (such as enclosures), it is important to know that sound waves will travel through air, water, wood, enclosures, etc., at different speeds.

- Transmission loss is the amount of energy lost as the sound wave travels through the enclosure walls. It is expressed in decibels. The openings in enclosures affect their efficiency, and the transmission loss for various building materials varies.

Absorption

- Noise control by absorption in room bounding surfaces is very limited in effectiveness. Seven to 10 dB is probably the maximum reduction that can be expected and, in most cases, 5 dB would be the best one could accomplish.

- Room absorption can only reduce the reverberant buildup in a room and therefore is not effective if the worker must be close to the noise source. Room absorption is most effective when the room has little or no sound absorbing material in it before treatment, and the room has multiple noise sources (four or more).

- A sabin is a measure of the sound absorption of a surface and is the equivalent of one square foot of a perfect absorptive surface.

- Noise reduction can be estimated by:

$$NR = 10 \log_{10} \frac{A_2}{A_1}$$

where:

NR = noise reduction in decibels

A_1 = total number of absorption units (sabins) in the room before treatment

A_2 = total number of absorption units (sabins) in the room after treatment

Absorption Coefficient

- Absorbing materials are generally soft and porous so that the sound wave can enter the material, but the material must have a high enough flow resistance to reduce the amplitude of the sound wave.

- If the material is too dense, sound will be reflected. If it is not dense enough, sound will pass through unchanged.

- Materials are rated for their ability to absorb sound by a sound absorption coefficient (α). This is the percent of incident sound that is absorbed in striking the material. Absorption coefficients vary with frequency.

- Example: If 55% of the incident sound energy is absorbed when it strikes the surface of a material, the α of that material would be 0.55.

Hearing Protection Devices

- There are many disadvantages to hearing protective devices. They do not reduce or eliminate the source of the noise. Their failure means immediate exposure. They can become ineffective without the knowledge of the user. And users must receive training on the devices' benefits, proper fitting, use, and care.

- Hearing protective devices serve as acoustic barriers that reduce the amount of sound energy transmitted through the ear canal to the receptors of the inner ear.

- Common types generally include earmuffs and inserts. Others include enclosures over the entire head and superaural protectors or earcaps.

Inserts

- Insert hearing protective devices are commonly called "earplugs." They are inexpensive and have limited service life. Earplugs are made of rubber, plastics, fine glass down, foam, etc.

- Earplug types include formable, custom-molded, and premolded.

Earmuffs

- Earmuffs consist of two cup- or dome-shaped devices that fit over the entire external ear, including the lobe, and a cushion or pad that seals against the side of the head. Efficiency is reduced when the fit is compromised by eyeglass wear.

Selection of Protector

- When selecting hearing protection devices, consider short-term or long-term use (i.e., impact noise vs. continuous noise for an entire shift).

- Examine carefully the manufacturer's reported test data. Check the noise reduction rating (NRR). Manufacturers must report the attenuation characteristics of the device over a range of frequencies.

- Refer to Appendix B of 29 CFR 1910.95 for using the NRR to determine whether a particular hearing protective device provides adequate protection within a given exposure environment.

Calculating Hearing Protector Attenuation

- Earmuffs, earplugs, and other noise reduction devices are usually given a noise reduction rating (NRR) by their manufacturer.

- OSHA requires that this rating be adjusted prior to evaluating its suitability on a work site.

- The formula for calculating the approximate field attenuation (FA) is outlined below. The value 7 is subtracted from the NRR to adjust for spectral uncertainty.

- The result is divided by 2 to produce a 50% safety factor. When two noise reduction devices are used, calculate the FA based on the most effective one, then add 5 for the second.

$$FA = \frac{NRR - 7}{2}$$

where:

FA = field attenuation in dBA
NRR = noise reduction rating (NRR) expressed in dBA

13 Industrial Ventilation

Ventilation is one of the most important engineering control techniques used by industrial hygienists, controlling workplace exposures with airflow. Ventilation is a very effective method for improving air quality in the workplace, preventing fire and explosion hazards, and control of heat and humidity for comfort. Airflow may be used for heating or cooling, removing a contaminant, diluting concentrations of a contaminant or supplying make-up air.

Local Exhaust Ventilation

LOCAL EXHAUST SYSTEMS

- A local exhaust ventilation system (LEV) is used to control an air contaminant at its source, as compared with a dilution ventilation system, which allows contaminants to be diluted in the workroom air.
- Local exhaust ventilation is preferred to dilution because it removes contaminants from the work area, it handles lower volume flow rates, and it is more energy efficient.
- Local exhaust ventilation often has the ability to scrub or clean the contaminant from the exhaust air.

Appropriate Use of Local Exhaust Systems

- A local exhaust ventilation system is usually appropriate for control when other, more cost-effective controls are not available.
- Local exhaust ventilation is effective when air sampling or employee complaints suggest that the air contaminant is a health hazard, a fire hazard, impairs productivity, or creates unacceptable comfort problems.
- Federal, state, or local building or consensus standards may require local exhaust systems (for example, 29 CFR 1910.94 calls for LEV for open surface tanks, grinding operations, etc.).
- Local exhaust ventilation systems may be used when improvements will be seen in productivity, housekeeping, employee comfort and morale, and the operation and maintenance of equipment.
- LEV is very effective when emission sources tend to be large, few, fixed, or widely dispersed. It is equally effective when emission sources are close to the breathing zones of employees and emission sources tend to vary with time.

Basic Terms and Formulas

Standard Conditions

- ACGIH defines standard conditions as 70° F, 29.92 in Hg, and dry air (407 inches of water (in. H_2O), 14.7 psi, 760 mmHg).

- The American Society of Heating, Refrigeration, and Air-Conditioning Engineers (ASHRAE) define standard air as 68° F, 29.92 in Hg, 50 percent relative humidity, and air density of 0.075lb/ft^3.

Pressure

- Air moves under the influence of differential pressures. A fan is commonly used to create a difference of pressure in duct systems.

Static Pressure

- Static pressure is the potential energy of the ventilation system. It is converted to kinetic energy (velocity pressure) or other energy types (losses in the form of heat, noise, and vibration).

- At sea level, standard static pressure is 14.7 psi, 29.92 in Hg, or 407 inches of water column.

Velocity Pressure

- Velocity pressure is directly related to the velocity of air, as described in Bernoulli's equation:

$$V = 4005\sqrt{VP}$$

where:

V = velocity of gases moving in the duct (ft/min, fpm, m/s, mps)
VP = velocity pressure of the gases moving in the duct (inches water gauge)

Velocity Pressure and Density Relationship

- Velocity pressure and density is the second basic relationship involving the velocity of gases that are flowing in a duct. It relates the velocity pressure and the density of the gases that are flowing through the duct.

$$V = 1096 \frac{\sqrt{VP}}{\sqrt{\rho}}$$

where:

V = velocity of the gases moving in the duct (fpm)
VP = velocity pressure of the gases moving in the duct (inches of water)
ρ = density of the gases flowing in the duct, measured in lbs/ft^3

Total Pressure

- The next three formulas are presented as a group because they constitute three "general rules" that will always apply to the gases (usually air) flowing through a duct.

- The first relationship states that static pressure (SP) and velocity pressure (VP) are related to total pressure (TP) through the following equation:

$$TP = SP + VP$$

	TP	SP	VP
Upstream*	-	-	+
Downstream**	+	+	+

*Upstream refers to pressure before the air reaches the fan.
**Downstream refers to pressure after the air reaches the fan.

Total Pressure Related to Losses in the Duct

- Air cannot travel through the duct without losing some energy. These loss factors are usually known or provided for in the form of a specific pressure loss or drop per unit length of the duct.

- The second relationship is expressed as:

$$TP_1 = TP_2 + \text{losses}$$

where:

TP_1 = total pressure at point 1 in the duct (inches of water)
TP_2 = total pressure at point 2 (downstream of point 1)
losses = pressure losses from various duct internal factors such as friction (inches of water)

Static Pressure and Velocity Pressure and Losses

- The third relationship is expressed as:

$$SP_1 + VP_1 = SP_2 + VP_2 + \text{losses}$$

where:

SP_n = static pressure at the nth point in the duct
VP_n = velocity pressure at the nth point in the duct
losses = Pressure losses from various duct internal factors (such as friction) that affect the flow of the gases in it

Flow Rate

- Volume flow rates are described by the conservation of mass formula.

$$Q = VA$$

where:

Q = volume flow rate (ft^3/min, cfm, scfm, acfm, m^3/s)
V = velocity (fpm, mps)
A = cross-sectional area of air flow (ft^2, m^2) $A = \pi r^2$

Components of a Local Exhaust System

LOCAL EXHAUST SYSTEM STRUCTURES

- Components of a local exhaust ventilation system include:
  - Hoods: Hoods capture, contain, or control the emission source.
  - Ductwork: Ducts and piping carry the contaminant to the air cleaner or to the outside environment for dilution.
  - Air cleaners: Air cleaners scrub, separate, remove, or filter the air contaminants for the exhaust air, usually to meet permit requirements. They may also change the air contaminant to a less hazardous substance.
  - Fans and motors: The fan generates static pressure and moves air.
  - Stacks: The stack assists in dispersing remaining air contaminants.

HOODS

- The hood in a local exhaust ventilation system is the point of air entry into the duct system. No local exhaust ventilation system can succeed unless the hood controls the contaminant. There are three basic types of hoods: capture, enclosing, and receiving (canopy).

Hood Entry Loss

- As air is pulled into a hood, it must overcome resistance, and will lose energy (i.e., suffer energy loss). This can be expressed as:

$$h_e = \frac{1 - C_e^2}{C_e^2} \bullet VP$$

where:

h_e = hood entry loss
C_e = coefficient of entry
VP = velocity pressure

Simple Hood Formula

- Hoods capture, contain, or receive contaminants before they have a chance to dilute into the workroom air. The hood converts static pressure in the duct (SP_h) into velocity pressure (VP) and hood entry losses (h_e).

- In simpler terms, we have to get the air moving. To do so means we will lose some energy in the process. This is the simple hood formula:

$$|SP_h| = VP + h_e \qquad \textbf{or} \qquad h_e = |SP_h| - VP$$

where:

h_e = hood entry loss (inches or mm water gauge)
VP = average velocity pressure in duct serving the hood (in or mm wg)
$|SP_h|$ = positive value of static pressure in duct serving hood (inches or mm water gauge)

Coefficient of Entry

- A measure of the effectiveness of hood in converting static pressure to velocity pressure is expressed through this efficiency formula.

$$C_e = \frac{\text{actual flow}}{\text{theoretical flow}} \qquad \textbf{or} \qquad C_e = \frac{\sqrt{VP}}{\sqrt{|SP_h|}}$$

where:

C_e = coefficient of entry (unitless)
VP = average velocity pressure in duct serving hood (in. or mm wg)
SP_h = positive value of SP_h (in. or mm wg)
h_e = hood entry loss

- In a perfect hood, we have no losses. When the coefficient of entry = 1, we have a perfect hood.

Hood Throat-Suction Equations

- The following two forms of this equation are known collectively as the hood throat-suction equations. These are used widely to determine volumetric flow rates of hoods.

$$Q = 4005\, A\sqrt{VP_d} \qquad \textbf{or} \qquad Q = 4005\, A\, C_e \sqrt{SP_h}$$

where:

Q = volumetric flow rate (cfm)
A = cross-sectional area of the hood opening (ft^2)
VP_d = velocity pressure in the hood duct (inches of water)
C_e = coefficient of entry (dimensionless)
SP_h = hood static pressure (inches of water)

Hood Entry Loss Factor

- The following equation is known as the hood entry loss factor equation.

$$F_h = \frac{h_e}{VP_h}$$

where:

F_h = hood entry loss factor (dimensionless)
h_e = hood entry losses (inches of water)
VP_h = velocity pressure in the hood (inches of water)

Compound Hood Equations

- The variations of the following formulae are known as the compound hood equations. An example of a compound hood is a side draft hood that almost always uses a slot.

- A slotted hood requires air to enter twice—first through the slot into the plenum, where velocity is reduced, and second from the plenum into a duct.

- The slot velocity should be twice the plenum velocity, or more. Slot hood entry loss factors are also compounded.

- For most hood and duct situations, the following two approximations hold:

$$H_{es} = 1.78[VP_s] \quad \text{and} \quad h_{ED} = 0.25[VP_d]$$

where:

H_{es} = hood slot entry loss
H_{ED} = hood duct entry loss

Hood Static Pressures in Compound Hoods

- The compound hood equations provide a vehicle for calculating hood static pressure for all conditions in which there are quantitative differences in the duct velocity and the hood slot velocity. There are two circumstances that are considered below.

Condition 1: The duct velocity is greater than the hood slot velocity.

$$SP_h = h_{ES} + h_{ED} + VP_d \text{ when } V_d > V_s$$

Condition 2: The hood slot velocity is greater than the duct velocity.

$$SP_h = h_{ES} + h_{ED} + VP_s \text{ when } V_s > V_d$$

where:

SP_h = hood static pressure
h_{ES} = hood slot entry loss
h_{ED} = hood duct entry loss
VP_d = duct velocity pressure
VP_s = slot velocity pressure
V_s = slot velocity
V_d = duct velocity

Air Flow into a Plain Round or Square Capture Hood

- With a freely suspended duct opening, air will move from all directions toward the face of a hood. Air will enter from the front and the back.
- The basic equation to measure air velocity is proportional to volume of air (Q) flowing into the duct in cfm or m^3/s, and inversely proportional to the square of the distance (x) from the opening.
- For round or square openings, the formula is represented as:

$$Q = V(10x^2 + A)$$

where:

V = centerline velocity (fpm)
x = distance from hood opening in feet (from hood face)
A = area of hood opening (ft^2)
Q = volume flow rate (cfm)

Air Flow into a Flanged Round or Square Capture Hood

- If a flange is placed around the duct opening, it will reduce entrance or turbulence loss by preventing the hood from drawing air from behind the hood face. This allows a larger portion of air entering the hood to come from the front of the duct, where the contaminant is produced.
- A large flange will increase useful airflow by 30 to 40% for the same total volume of air handled.
- This basic equation is expressed as:

$$Q = 0.75V(10x^2 + A)$$

where:

Q = air flow (cfm)
V = capture velocity
x = distance from hood opening (feet)
A = area of hood opening (ft^2)

Air Flow into a Free Standing Slotted Capture Hood

- The aerodynamic behavior of long, narrow hood openings differs from round or square hood openings. These are commonly used to provide uniform air distribution across a large surface (e.g., dip tanks).

- The relationship is expressed as:

$$Q = 3.7\,l\,V\,x$$

where:

Q	= air flow (cfm)
l	= length of slot (feet)
V	= capture velocity (usually 100 fpm)
x	= distance from hood opening (feet)

- The absence of the slot width is intentional in this formula. Treat a multiple slotted hood as a plain hood, using length and width of the slots to determine the hood size.

Air Flow into a Flanged Slotted Capture Hood

- Similar to a plain hood, adding a flange to a slotted hood increases its efficiency.

- This relationship is expressed as:

$$Q = 2.6\,l\,V\,x$$

where:

Q	= air flow (cfm)
l	= length of slot (feet)
V	= capture velocity
x	= distance from hood opening (feet)

Enclosures

- Most control velocities at the point of generation are relatively low and can be nullified by drafts from cross breezes or nearby moving equipment.

- Cross drafts can be eliminated by using baffles, side shields, booths, and other semi-enclosures.

- Full enclosure acts differently and more efficiently than a capture hood because it completely envelops the operation. With a capture hood, the air velocity at the point of emission must be high enough to transport the contaminant through free space to the hood and duct without diversion of a cross draft.

Receiving Hoods (Canopy Hoods)

- Receiving hoods (or canopy hoods) are seldom successful as capture hoods since:
  - Air velocity drops off tremendously (in proportion to the square of the distance) between the canopy and source.
  - A greater volume of airflow is required for contaminant control.
  - Control may be lost due to room air cross drafts.
  - The canopy type of hood often forces the employee to work beneath it, that is, between the point of contaminant emission and the hood.
- Receiving hoods are mostly used where rising hot air carries the contaminant into the exhaust system.
- The amount of air required for a canopy hood can be approximated by the following equation:

$$Q = 1.4PVx$$

where:

Q = rate of airflow (cfm)
P = perimeter length of hood (ft)
x = height of canopy above tank (or distance from hood opening) (ft)
V = desired control velocity (fpm)

Hood Design

- Good exhaust hood design should include:
  - Enclosing the operation as much as possible to reduce the rate of airflow needed to control the contaminant and prevent cross drafts.
  - Locating the hood so that the contaminant is moved away from the breathing zone of the operator.
  - Locating and shaping the hood so that the initial velocity of the contaminant will throw it into the hood opening.
  - Capturing solvent vapors that are heavier than air rather than attempting to collect them at the floor level.
  - Designing the hood so that it will not interfere with the worker.

DUCTS

- Ducts carry air to an air cleaner or to the outdoors. When air passes through any duct or pipe, friction must be overcome. The amount of this friction loss must be calculated before the system is installed so that the proper size fan and motor can be purchased.

Duct Design

- Careful duct design:
  - Holds power consumption to a minimum.
  - Maintains proper transport velocity so the contaminant, if it is a dust or fume, does not settle out and plug the duct.
  - Keeps the system balanced at all times.

Multiple Ducts

- Local exhaust systems with multiple ducts may pose problems. The static pressure between a junction point and the face of hood A must be the same as between the junction and the face of hood B. For proper volume flow rate of each hood, balance must be achieved.

- Balance can be achieved by sizing the ductwork and selecting fittings that will give approximately equal pressure drops in the two branches.

- The advantages of a properly balanced system include:
  - If velocities are initially chosen properly, ductwork will not become plugged.
  - The system cannot be tampered with.
  - Erosion will be reduced because no buildup is caused by damper obstruction.

- The advantages of using dampers in each branch include:
  - Corrects improperly estimated exhaust volumes.
  - Provides flexibility for future changes to the system.

Duct Velocity Terminology

- Static pressure (SP): Static pressure is created by the fan and is the energy source of the system. Static pressure is converted to velocity pressure and heat in the form of friction loss.

- Velocity pressure (VP): Moving air creates velocity pressure. Converting static pressure to air movement at the duct serving the hood creates velocity pressure.

- Duct friction loss: For the same velocity of air, narrow diameter ducts have higher friction loss than large diameter ducts.

- Other losses: Elbows and tapered sections where a branch enters a main branch or duct adds losses. The sharper the bend and more abrupt the taper in transition pieces, the higher the loss.

Air Flow in Ducts

- As the area within a duct changes, velocity changes to maintain a constant airflow (Q).

- This airflow relationship is expressed as follows:

$$V_1A_1 = V_2A_2$$

where:

V_1 = velocity of first section of the duct
A_1 = area of the first section in square feet
V_2 = velocity of second section
A_2 = area of second section

AIR CLEANERS

- There are two broad categories of industrial air cleaners. These include:
  - Industrial air cleaners that remove airborne contaminants that would otherwise pollute the surroundings.
  - Air cleaners that handle relatively high rates of airflow at low static pressure and that are associated with heating, ventilating, and air conditioning systems.

- Industrial air cleaners include settling chambers, wet and dry centrifugals, wet and dry dynamic precipitators, electrostatic precipitators, wet-packed scrubbers, washers, and fabric filters.

CENTRIFUGAL FANS

- Centrifugal fans are modifications of the wheel type and can be used where static pressure is medium to high (2 or more in. water gauge (wg)).

- Centrifugal fans include straight bladed radial fans, backward curved bladed fans, and forward bladed fans.

- Straight or forward-curved blades demand more power as airflow increases. Backward-curved blades have a power demand that reaches a maximum. The driving motor cannot be overloaded at the given speed.

Straight Bladed Radial Fans

- Straight bladed radial fans are the most common. This wheel type is the workhorse of fans in industrial ventilation.

- Steel or cast iron blades do not clog easily and withstand considerable abrasion. This fan type has medium tip speed, a high noise factor, and low mechanical efficiency.

- Straight bladed radial fans are often used in woodworking shops where lint, chips, etc., pass through blades.

Backward-Curved Fans

- Backward-curved blades permit higher tip speed and greater fan efficiency. Material will build up on blades because of tilt of blade and because it has an air cleaner ahead of it. There is a higher noise factor.
- Its high efficiency and non-overloading qualities make it a choice for large-volume exhaust.

Forward Bladed Fans

- Forward bladed fans are tipped in the direction of rotation, have low space requirements and low tip speed. There is little noise, and they are popular in small heating and air conditioning systems.

AXIAL FLOW FANS

- Axial flow fans are similar to modifications of the familiar desk fan or propeller fan.
- Propeller fans are average cooling fans that move large volumes of air against negligible resistance. They are usually mounted on pedestals for cooling individuals. No ducts are connected.
- Narrow-blade propeller fans are often used in spray booth exhausts. Duct connections must be kept to a minimum. These fans are sensitive to added resistance.
- Tube-axial fans are improved versions of the propeller fan. They are mounted in short sections of ducts to which piping can be attached.
- Vane-axial fans have guide vanes in short pipe sections that straighten the airflow. They operate against low static pressure (up to 4 in. wg).

Fan Speed and Air Flow

- Air volume (Q) and fan speed (RPM) are linearly related. This relationship is expressed as:

$$Q_N = Q_I \times \frac{\text{New RPM}}{\text{Initial RPM}} \quad \textbf{or} \quad \frac{Q_1}{Q_2} = \frac{RPM_1}{RPM_2}$$

where:

Q_N = new or final air flow (cfm)
Q_I = initial air flow (cfm)
RPM = revolutions per minute

Fan Speed and Fan Static Pressure

- Static pressure (SP) and fan speed (RPM) are related through a second-power relationship. This relationship is expressed as:

$$FSP_N = FSP_I \times \frac{(\text{New RPM})^2}{(\text{Initial RPM})^2} \quad \textbf{or} \quad \frac{SP_1}{SP_2} = \frac{(RPM_1)^2}{(RPM_2)^2}$$

where:

FSP_N = new fan static pressure (in. wg)
FSP_I = initial fan static pressure (in. wg)
RPM = revolutions per minute

Fan Speed and Horsepower

- Horsepower (BHP) and fan speed (RPM) are related through a third power relationship. This relationship is expressed as:

$$HP_N = HP_I \times \frac{(\text{New RPM})^3}{(\text{Initial RPM})^3} \quad \textbf{or} \quad \frac{HP_1}{HP_2} = \frac{(RPM_1)^3}{(RPM_2)^3}$$

where:

HP_N = new or final horsepower, bhp
HP_I = initial horsepower, bhp
RPM = revolutions per minute

Determining Horsepower Needs

- The general term "horsepower" may refer to brake horsepower, rated horsepower, or motor horsepower.
- Fans are not perfectly efficient. When a fan moves a specified volume (cfm) against a specified fan total pressure (TPf), it consumes 20–50 percent of the input power, depending on the fan.
- Brake horsepower takes into account the fan inefficiencies, shaft and bearing losses in the fan, and other fan losses.

Air Horsepower (AHP)

- Air horsepower refers to the minimum amount of power needed to move a volume of air against the fan total pressure. This relationship is expressed as:

$$AHP = \frac{Q\,(TPf)}{6356}$$

where:

Q = volume of air (cfm)
TPf = fan total pressure
AHP = air horsepower

Motor Horsepower

- Motor horsepower (MHP) refers to brake horsepower plus any power required to account for drive losses, slippage, and pulley losses. This relationship is expressed as follows:

$$MHP = bhp \bullet Kd$$

where:

MHP = motor horsepower
bhp = brake horsepower
Kd = fan drive (obtained from manufacturer)
Kd = 1.15 pulley drive
Kd = 1.05 direct drive

Brake Horsepower

- Brake horsepower refers to the actual horsepower required to operate the fan so that it fulfills its job of moving the specified cfm against the specified TPf.

- This relationship is expressed as:

$$BHP = \frac{Q\,(TPf)}{6356(ME)}$$

where:

BHP = brake horsepower
Q = volume of air (cfm)
TPf = fan total pressure
ME = mechanical efficiency

- Mechanical efficiency is the efficiency factor determined by the fan manufacturer. If it is not available, use the following:

ME = 0.65 for radial fan
ME = 0.70 for BC or FC
ME = 0.75 for axial fan

Rated Horsepower

- Rated horsepower is normally the horsepower indicated on the motor's nameplate. A 5 hp motor means that the motor is designed to draw up to five horsepower's worth of current from the electrical lines serving it. It doesn't mean that the motor will always be supplying five horsepower.

Calculations Involving the Diameter of Fans

- The following equation relates the air discharge volume being produced by a fan to its diameter. This relationship assumes that the fan being examined will be evaluated for different diameters and/or different densities of the gases and fluids.

- The formula for calculations involving the diameter of fans is expressed as follows:

$$Q_2 = Q_1 \bullet \frac{(Size_2)^3 \bullet (RPM_2)}{(Size_1)^3 \bullet (RPM_1)}$$

where:

Q_2 = new or final air discharge volume (cfm)
Q_1 = initial air discharge volume (cfm)
RPM_n = revolutions per minute of the nith fan
$Size_n$ = diameter of the nth fan, measured in inches

Static Pressure and Fan Diameters

- The following equation relates static discharge pressure being developed by a fan to its diameter:

$$SP_2 = SP_1 \bullet \frac{(Size_2)^2}{(Size_1)^2} \bullet \frac{(RPM_2)^2}{(RPM_1)^2} \bullet \frac{(\rho_2)}{(\rho_1)}$$

where:

SP_2 = New or final static discharge pressure of a fan (inches of water)
SP_1 = Initial static discharge pressure of a fan (inches of water)
$Size_n$ = Diameter of the nth fan, measured in inches
RPM_n = Revolutions per minute of the nth fan
ρ_n = the nth density of the gases flowing through the fan (lbs/ft^3)

Brake Horsepower and Fan Diameter

- The following equation relates the brake horsepower required operating fans of various diameters.

$$BHP_2 = BHP_1 \bullet \frac{(Size_2)^5}{(Size_1)^5} \bullet \frac{(RPM_2)^3}{(RPM_1)^3} \bullet \frac{(\rho_2)}{(\rho_1)}$$

where:

BHP_2 = New or final horsepower required for a fan of ith diameter (bhp)

BHP_n = Initial or original horsepower required for a fan of nth diameter

$Size_n$ = Diameter of the nth fan measured in inches

RPM_n = revolutions per minute or rotational speed of the fan (rpm)

ρ_n = the nth density of the gas flowing through the fan (lbs/ft³)

Fan Static Pressure

- Another relationship between fan static pressure and its relationship to velocity pressure is expressed as:

$$FSP = SP_{out} - SP_{in} - VP_{in}$$

where:

FSP = fan static discharge pressure (inches of water)

SP_{out} = fan outlet static pressure (inches of water)

SP_{in} = fan inlet static pressure (inches of water)

Vp_{in} = fan inlet velocity pressure (inches of water)

Total Fan Pressure

- The fan total pressure equation is a straight forward relationship expressed as:

$$Tp_{fan} = TP_{out} - TP_{in}$$

where:

TP_{fan} = fan total pressure (inches of water)

TP_{out} = fan total outlet pressure (inches of water)

TP_{in} = fan total inlet pressure (inches of water)

General Ventilation Systems

APPROPRIATE USES OF DILUTION VENTILATION

- Dilution ventilation or general ventilation systems in the industrial environment work well when other more cost-effective controls are unavailable.

- Use dilution ventilation when:
  - Air contaminants are of relatively low toxicity.
  - The air contaminants consist of gases, vapors, or small aerosols rather than dusts or fumes.
  - Emissions occur uniformly in space and time.
  - Emission sources are widely dispersed and are not close to employee breathing zones.
  - The facility is located in a moderate climate and the air used for dilution is not contaminated.
  - The dilution air must be capable of maintaining appropriate temperature and humidity.

Properties of Air Mixing

- Contaminants may mix uniformly in air. However, some contaminants may be lighter or heavier than air and will not uniformly mix with air.

- A mixing factor must be selected to account for poor or incomplete mixing. The mixing factor (K) is available in the ACGIH's *Industrial Ventilation Manual.*

- Termed ventilation efficiency or ventilation effectiveness, the mixing factor is stated mathematically as:

$$K_{mix} = \frac{\text{Actual } Q_{dilution} \text{ required to achieve the acceptable concentration}}{\text{Ideal } Q_{dilution} \text{ required to achieve the acceptable concentration}}$$

where:

$Q_{dilution}$ = volume flow rate of dilution air

K_{mix} = K factor (when greater than 3, use local exhaust ventilation)

Rate of Evaporation

- Evaporation or emission rate is generally exponential over time, but can be approximated linearly. It is known that a given volume of vapor forms from a given weight of solvent evaporated.

Dilution Ventilation for Gases and Vapors

- Use this generalized formula for calculating the airflow necessary to dilute a gaseous or vapor emission to "safe" levels.

- This should be used for low toxicity compounds and when the generation rate is constant. The K factor as discussed above represents a safety factor. The common factor used for gases and vapors is 10.

❑ The formula is expressed as:

$$Q = \frac{KG}{C} \bullet 10^6$$

whcre:

Q = quantity of air (cfm)
K = constant (usually 10 for gases and vapors)
G = generation rate (cfm)
C = PEL or TLV

Dilution Volume Flow Rate (lbs)

❑ When working with pounds of a contaminant, calculate the dilution volume flow rate as follows:

$$Q_{dilution} = \frac{387^{*} \text{ x lbs. x } 10^6 \text{ x } K_{mix}}{MW \text{ x } t \text{ x } C \text{ x } d}$$

where:

$Q_{dilution}$ = volume flow rate of dilution air (cfm)
t = time (minutes)
MW = molecular weight of material emitted
lbs = amount evaporated (lbs)
C = the acceptable exposure concentration (ppm)
K_{mix} = mixing factor to account for incomplete or poor mixing in the space
d = density correction factor to account for nonstandard conditions (d = 1 near sea level and normal room temperature)

*A conversion used to solve this problem is: 387 ft^3 = 1lb-mole.

Dilution Ventilation Volume—Pints Per Hour

❑ When contaminants are released into a workplace in pints per hour, calculate the dilution volume flow rates as follows:

$$Q_{dilution} = \frac{403 \text{ x } SG \text{ x } E \text{ x } 10^6 \text{ x } K_{mix}}{MW \text{ x } C}$$

where:

$Q_{dilution}$ = volume flow rate of dilution air (cfm)
SG = specific gravity
MW = molecular weight of material emitted
E = emission rate (pints per minute)
C = PEL or TLV
K_{mix} = mixing factor to account for incomplete or poor mixing in the space

Dilution Ventilation for Fire and Explosion Hazards

- For fire and explosion hazards, use this revised formula, which incorporates a safety factor and an adjustment for temperature (B).

$$Q_{dilution} = \frac{403 \times SG \times E \times S_F \times 100}{MW \times LEL \times B}$$

where:

$Q_{dilution}$ = volume flow rate of dilution air (cfm)
S_F = safety factor 25% of LEL use 4, 10% of LEL use 10
SG = specific gravity
MW = molecular weight of material emitted
E = emission rate (pints per minute)
B = if temp ≥ 250° F, B = 0.7, if temp is < 250° F, B = 1
LEL = lower explosive limit (%)

Dilution Ventilation—Steady State Conditions

- Volume of room is important in determining how much air is necessary to bring the level down to safe exposure levels. It assumes the contaminant level of dilution ventilation air is zero.

- This formula determines the time necessary to reduce a static air concentration to a desired level and assumes perfect air mixing.

$$t = \frac{V_r}{Q} \, ln \frac{(C_i)}{(C)}$$

where:

t = time (minutes)
C = desired concentration (TLV or PEL)
C_i = initial concentration
V_r = volume of area/room (ft³)
Q = air flow into room (cfm)

Formula for Determining Time to Reduce Contaminants

- An alternative formula can be used to determine the time necessary to reduce an air concentration to a desired level.

$$t = \text{Log} \frac{C^{-2.303}}{C_o} \bullet \frac{P}{Q}$$

where:

t	=	time (minutes)
C	=	new or desired concentration (TLV or PEL)
C_o	=	initial or original concentration (ppm)
P	=	volume of the room or area (ft^3)
Q	=	air flow into room (cfm)

Purge -Dilution Equation

- The following equation, known as the purge-dilution equation, is the most basic and fundamental relationship available for determining the various parameters associated with reducing the concentration of the vapor of any volatile material in a closed space or room.

$$C = C_0 e^{-[V\ removed/V\ room]}$$

where:

C	=	ending concentration of the vapor in the closed space or room (ppm)
C_0	=	initial concentration of the vapor in the closed space or room that is to be reduced by purging (ppm)
$V_{removed}$	=	air volume that has been withdrawn from the closed space or room, measured in any suitable volumetric units (ft^3)
V_{room}	=	volume of the room, measured in the same volumetric units (ft^3)

Purge -Build Up Equation

- The concentration of a gas or vapor at any time can be expressed by a differential material balance, which, when integrated, provides a rational basis for relating ventilation to the generation and removal rates of contaminants.

- Starting with a fundamental material balance, assuming no contaminant is in the supply air, then:

  Rate of Accumulation = Rate of Generation - Rate of Removal

- The relationship is expressed as follows:

$$ln \frac{(G - Q'C_2)}{(G - Q'C_1)} = \frac{-Q'(t_2 - t_1)}{V_{room}}$$

where:

G = generation rate of contaminant

$Q' = \frac{Q}{K}$ = effective rate of ventilation, corrected for incomplete mixing (cfm)

K = design distribution constant, allowing for incomplete mixing

V = volume of the room or enclosure (ft^3)

t_1 = initial time (minutes)

t_2 = elapsed time (minutes)

C_2 = the new concentration of the contaminant

C_1 = the original concentration at time t

Air Changes Per Hour

- In assessing indoor air quality, it is often necessary to determine the number of air changes per hour (N) for a given room or space.
- Outlined below is a simple formula:

$$N = \frac{Q \bullet 60}{Vol}$$

where:

N = air changes per hour

Q = air flow (cfm)

Vol = room volume (ft^3)

Evaluating Ventilation Systems

PURPOSE OF EVALUATING VENTILATION SYSTEMS

- Evaluate ventilation systems to:
  - Assure adequacy of design and performance.
  - Assure that system performance is maintained.
  - Determine the feasibility for expanding the system.
  - Establish improved design parameters for new systems.
  - Assure compliance with federal, state, or local regulations.

EQUIPMENT AND METHODS USED

- Several types of equipment are used in evaluating ventilation systems, including smoke tubes, rotating vane anemometers, swinging vane anemometers, heated thermocouple anemometers, heated wire anemometers, and pitot tubes.

Smoke Tubes

- A smoke tube is a descriptive term applied to a glass tube containing titanium tetrachloride adsorbed on a granular medium. The smoke tube operates in the following manner:
  - The ends of the tube are broken and air passes through the tube.
  - Moisture in the air reacts with the reagent to form hydrochloric acid "smoke."
  - The tubes are held close to the hood initially, and then gradually away from the hood entry to observe the sphere of containment.

- Limitations of smoke tubes include they following:
  - Smoke tubes provide strictly qualitative analysis.
  - They do not provide any information in terms of design or performance specifications.
  - The titanium tetrachloride used in tubes is an irritant.

Rotating Vane Anemometers

- Rotating vane anemometers measure air velocity and flow. These are best suited for determining air velocities and estimating air flow through large openings, such as mine shafts and air supply and discharge grills.

- Their optimum range is between 100 and 3000 fpm.

- They can only be used in clean air, require calibration, and must be handled with care.

- Principles of operation:
  - Comprised of a vane or propeller on a shaft connected to gears.
  - Air movement causes the vane to rotate.
  - Revolutions of the vane turn the gears that register the revolutions on the dial of the instrument as linear feet.
  - Readings are usually taken over one-minute periods.

Swinging Vane Anemometers

- Swinging vane anemometers measure air velocity. They are portable and used for measuring velocities of exhaust or supply for all types of local exhaust systems.

- Fittings are available for measuring static pressures, as well as a wide range of velocities from 25 to 24,000 fpm.

- Their optimum range of performance is 100 to 10,000 fpm.

- When air temperatures vary from standard pressure and temperature, the anemometer must be corrected.

- Principle of operation: Indicates air velocity as a function of the pressure exerted by the air stream against a spring loaded swinging vane.

Heated Thermocouple Anemometers

- Heated thermocouple anemometers are other air velocity measuring devices. These units are portable, which makes them applicable to fieldwork. The range of air velocities includes 10 to 10,000 fpm. Periodic calibration is required.

- Heated thermocouple anemometers may be used in determining non-directional velocities, such as general room air movement. They can also be used for entry or exit velocities of hoods, slots, grills, etc.

- Limitations: The integrity of the probe must be maintained. Heavy dust or corrosive materials, as well as mechanical shock, could damage the delicate wires in the probe.

- Principle of operation:
  - Air moving past a heated object removes heat.
  - The amount of heat removed is proportional to the quantity of air passing, which is a function of velocity.
  - A change in airflow causes a change in temperature of the thermocouples, resulting in a change in the direct current output.
  - Place probe perpendicular to the direction of airflow and record the measurements.

Heated Wire Anemometers

- Another velocity meter frequently used is the heated wire anemometer.

- These are not amenable to fieldwork because of the fragile thermometers and the amount of time required for the thermometers to reach equilibrium. They are primarily used in laboratories for the calibration of air samplers.

- Principle of operation: Similar to the heated thermocouple anemometers, except that two thermometers are used instead of thermocouples.

Measurements Within a System

- Rather than measuring the velocity of air going into exhaust hoods, measurements can be made inside the ductwork leading to the point of entry or discharge. Such measurements are made to determine static pressure drops associated with hood entries, ducts, and across air cleaners, as well as velocity pressures.

Pitot Tube Velocity Pressure Measurements

- The pitot tube is the standard instrument for measuring the velocity of air in ducts. Use pitot tubes when determining velocity pressure contours inside of ducts.

- Characteristics of pitot tubes include the following:
  - Pitot tubes are considered a primary standard and require no calibration.
  - Pitot tubes are resistant to corrosion and can withstand high temperatures.
  - They can be used in the field.
  - Their range is best between 600–800 fpm.
  - Pitot tubes have fewer limitations than other air velocity measuring devices.
  - Pitot tubes can be used in corrosive environments with temperature variations.
  - Their openings can become clogged with particulates.
  - They cannot be used for low velocities (< 600 fpm).

- Principles of operation:
  - The pitot tube consists of two concentric tubes.
  - The opening of the inner tube is axial to the flow and measures total pressure.
  - The larger tube with circumferential openings measures static pressure.
  - Measurements are obtained by connecting the static and total pressure taps to a manometer. Inclined manometers are usually used since they increase precision and accuracy for velocities below 2000 fpm.
  - The difference between the total pressure and static pressure is the velocity pressure.

- Location of the measurement in a duct is important for validity:
  - Airflow is not uniform in the cross section of a duct.
  - This is specially true near elbows, entries, etc.
  - Measurement must be taken at least 8.5 diameters of straight run downstream or 1.5 diameters upstream from interference.
  - Once the location is selected, a pitot traverse can be conducted.

Calculations Involving Air Flow Balancing at a Duct Junction

- When evaluating ventilation systems, balancing a system is critical.

- The process of balancing a duct junction is logical in its development and in the solution path it implies.

- In this procedure, consideration is given to the volumetric flow rate in each of two ducts that join to form a single larger duct.

- This relationship can be expressed as:

$$R = \frac{SP_{greater}}{SP_{lesser}}$$

where:

R = the junction balance ratio, which is a dimensionless number

$SP_{greater}$ = the duct static pressure in the duct that carries the greater flow volume into the duct junction (inches of water)

SP_{lesser} = the duct static pressure in the duct carrying the lesser flow volume into the duct junction (inches of water)

A Balanced System

- The first scenario to consider is when $R < 1.05$. This junction is considered balanced.

- In other terms, when the difference in static pressure requirements at the junction is less than 5%, ignore the difference.

When Corrections Can Be Made

- When $1.05 \leq R \leq 1.20$, increase volumetric flow rate in the lesser volume branch from Q_{design} to Q_{cor} according to the following formula:

$$Q_{cor} = Q_{design} \frac{\sqrt{SP_{greater}}}{\sqrt{Sp_{lesser}}}$$

where:

Q_{cor} = new or corrected volumetric flow rate

Q_{design} = initial or earlier designed volumetric flow rate

$SP_{greater}$ = duct static pressure in the duct that carries the greater flow volume into the duct junction (inches of water)

Sp_{lesser} = duct static pressure in the duct that carries the lesser flow volume into the duct junction (inches of water)

- When the difference in static pressure requirements at the junction ranges from 6 to 20%, correct the volume in the non-governing duct.

When the System Cannot Be Balanced

- When R > 1.20, the duct junction is so imbalanced that the entire ventilation system must be completely redesigned.
- When the difference in static pressure requirements at the junction exceeds 20%, resize the ductwork, or provide additional static pressure loss.

Sample Problems

Sample Problem 1

- What dilution volume flow rate should be used if 14 pounds of toluene are uniformly evaporated from a process during an eight-hour period? Assume that $C_a = 10$ ppm, $K_{mix} = 1.5$, $d = 1$, and $MW = 92.1$

$$Q_{dilution} = \frac{387 \text{ x lbs x } 10^6 \text{ x } K_{mix}}{MW \text{ x t x } C_a \text{ x d}}$$

$$= \frac{387 \text{ x } 14 \text{ x } 10^6 \text{ x } 1.5}{92.1 \text{ x } 480 \text{ x } 10 \text{ x } 1}$$

$$= 18{,}000 \text{ cfm (rounded)}$$

Sample Problem 2

- A local exhaust system to control aluminum oxide fumes requires 300 cfm of airflow with a design transport velocity of 2000 fpm. Choose the best available duct size.

$$Q = VA$$

$$A = \frac{300}{2000} = 0.15 \text{ ft}^2$$

duct diameter = 5.2 inches

Sample Problem 3

- The velocity pressure of 0.6 inches of water corresponds to what velocity in a six-inch duct?

$$V = 4005\sqrt{VP}$$

$$= 4005\sqrt{0.6}$$

$$= 3102 \text{ fpm}$$

Sample Problem 4

- If a velocity pressure of 0.6 inches of water and a total pressure of -0.2 inches of water were measured in a duct on the suction side of the fan, what static pressure would be expected?

$$TP = SP + VP$$
$$-0.2 = SP + 0.6$$
$$SP = -0.8 \text{ inches of water}$$

Sample Problem 5

- A vapor source is located at a distance of three feet from the face of a 3-ft x 3-ft unflanged square hood. Assume a duct velocity of 1800 fpm and a capture velocity of 90 fpm. Calculate the exhaust flow rate and the centerline air velocity at one foot from the hood face.

$$Q = V(10x^2 + A)$$
$$= 90\,[10(3)^2 + 9]$$
$$= 8910 \text{ cfm}$$

$$V = \frac{Q}{10x^2 + A}$$
$$= \frac{8910}{10(1)^2 + 9}$$
$$= 469 \text{ fpm}$$

Sample Problem 6

- For the same hood with a six inch wide flange, calculate the exhaust flow rate and the centerline air velocity at one foot from the hood face.

$$Q = 0.75\,(V)\,(10x^2 + A)$$
$$= 0.75\,(90)\,[10(3)^2 + 9]$$
$$= 6682.5 \text{ cfm}$$

$$V = \frac{Q}{(0.75)(10x^2 + A)}$$
$$= \frac{6682.5}{0.75\,(10\,(1)^2 + 9)}$$
$$= 469 \text{ fpm}$$

Sample Problem 7

- The entry loss for a simple inlet is measured to be 10 in. wg. The velocity pressure in the duct is 20 in. wg. Calculate the hood static pressure.

$$SP_h = VP_d + h_e$$
$$= 20 + 10$$
$$= 30 \text{ in. wg}$$

Sample Problem 8

- The coefficient of entry for a particular hood is 0.73. The velocity pressure is 25 in. wg. Calculate the hood entry loss.

$$h_e = \frac{1 - C_e^2}{C_e^2} \times VP$$

$$= \frac{1 - (0.73)^2}{(0.73)^2} \times 25$$

$$= 21.9 \text{ in. } H_2O$$

Sample Problem 9

- A fan has been selected to deliver 35,000 cfm against a 7.5" SP. The fan operates at 1200 rpm and requires 60.0 bhp. After installation, it is desired to increase the fan's output by 20%. At what rpm must the fan be operated to achieve this? Is a new SP developed? Is a new bhp required?

$$CFM = RPM \qquad 1200 \bullet 1.2 = 1440 \text{ rpm}$$

$$SP = RPM^2 \qquad \frac{(1440)^2}{(1200)^2} \bullet 7.5 = 10.8''$$

$$BHP = RPM^3 \qquad \frac{(1440)^3}{(1200)^3} \bullet 60 = 103.7 \text{ bhp}$$

14 Statistics in Industrial Hygiene

When evaluating the workplace, industrial hygienists must ensure that representative air samples are collected. Collecting air samples with minimum variations and errors is critical to inferring conclusions about air sampling results. Industrial hygienists must understand how to prevent errors and account for them when they do occur using statistical methods.

Air Sampling Errors, Accuracy, and Precision

SAMPLING ERRORS

- Industrial hygiene air monitoring evaluates the work environment, but it only measures part of the environment and, from the measurements taken, infers conclusions about the whole.
- In all sampling methodologies, there are both systematic and random errors to consider that can affect the interpretation of the results and the final judgment about the work environment as a whole.
- Any exposure average calculated from air-sampling measurements is only an estimate of the true exposure. It is important to know where errors can occur to prevent their occurrence and/or account for them when interpreting results.

Accuracy

- Accuracy is the relationship between a measured value and the true value. For a measure to be accurate, it must be close to the true value.
- Accuracy is affected by controllable sources of error, termed determinate or systematic errors. These include method error (e.g., incorrect calculations, use of contaminated reagents), personal error (e.g., personal carelessness), and instrument error (e.g., poorly calibrated equipment).

Precision

- Precision is the degree of agreement among results obtained by repeated measurements under the same conditions and under a given set of parameters.
- Precision is affected in indeterminate or random errors. These cannot be controlled. These include sampling equipment variations (e.g., random pump flow fluctuations), analytical method fluctuations (e.g., variation in reagent addition or instrument response), and intra- or inter-day concentration fluctuations.

- These factors cause variability among the sample results. Statistical techniques are used to account for random error. Increasing the number of samples taken minimizes the effect of random error.

SAMPLING ERRORS AND OSHA COMPLIANCE

- To ensure compliance or violation, OSHA compliance officers use one-sided confidence limits (upper and lower confidence limits, UCL and LCL) whenever sampling is performed. (*OSHA Technical Manual*, 4th edition).

- This practice recognizes that the sample measured on the employee is rarely the same as the "true" exposure because of sampling and analytical errors (SAEs).

Upper Confidence Limits (UCL) and Lower Confidence Limits (LCL)

- The upper confidence limit (UCL) and lower confidence limit (LCL) incorporate these error factors statistically in order to obtain the lowest (LCL) and the highest (UCL) value that the true exposure could be, within a 95 percent confidence interval.

- The UCL and LCL are one-sided limits because they are used by both OSHA and employers to ensure that the true exposure lies on one side of the OSHA permissible exposure limit (PEL), either above or below it.

Coefficient of Variation

- The coefficient of variation (CV_T) may be found in the *NIOSH Manual of Analytical Methods*, or it may be computed.

- To calculate the UCL and LCL, first compute the coefficient of variation (CV_T) for each analytical method.

$$CV_T = 100 \bullet \frac{sd}{\mu}$$

where:

sd = standard deviation
μ = mean (or analytical result)
100 = factor to convert from fraction to percent

Sampling and Analytical Errors (SAE) Derived from the CV_T

- Sampling and analytical errors (SAEs) are often listed in OSHA report forms but can be derived from the CV_T.

- Use the following formula for deriving the SAE from the CV_T:

$$SAE = 1.645\ (CV_T)$$

where:

$$CV_T = CV_S + CV_A$$

or the coefficient of variation of the sampling method (CV_s) plus the coefficient variation of the analytical method (CV_A)

Lower Confidence Limit (LCL) Formula

- Use the following formula for calculating the lower confidence limit at the 95% confidence level.

$$LCL(95\%) = \frac{\mu}{PEL} - 1.645\ CV_T$$

where:

LCL(95%) = lower confidence limit at the 95th percentile
μ = mean (average of the airborne concentration)
CV_T = CV total or coefficient of variation including sampling error
n = number of data points to determine μ
1.645 = appropriate factor from large sample statistics (obtained from a statistical Z table)

Upper Confidence Limit (UCL) Formula

- In a similar fashion, use the following formula for calculating the upper confidence limit.

$$UCL(95\%) = \frac{\mu}{PEL} + 1.645\ CV_T$$

where:

UCL(95%) = upper confidence limit at the 95th percentile
μ = mean (average of the airborne concentration)
CV_T = CV total or coefficient of variation including sampling error
n = number of data points to determine μ
1.645 = appropriate factor from large sample statistics (obtained from a statistical Z table)

Classifying Exposures

THREE CATEGORIES OF EXPOSURE

- One-sided confidence limits can be used to classify the measured exposure into one of three categories: compliance, possible overexposure, and violation.

- ❑ Compliance: If the measured results do not exceed the standard and the UCL also does not exceed the standard, we can be 95% confident that the measured exposure is in compliance.

- ❑ Violation: If the measured exposure exceeds the PEL and the LCL of that exposure also exceeds the PEL, we can be 95% confident that the measured exposure is in noncompliance, and a violation is established.

- ❑ Possible overexposure: If the measured exposure does not exceed the PEL, but the UCL of that exposure does exceed the PEL, we cannot be 95% confident that the employee's exposure is in compliance. If the measured exposure exceeds the PEL, but the LCL of that exposure is below the PEL, we cannot be 95% confident that the measured exposure is in noncompliance. In both of these cases, the measured exposure can be termed a "possible overexposure." A violation is not established in the "possible overexposure" region.

THREE SAMPLE TYPES

- ❑ OSHA uses simplified versions of the LCL and UCL formulas and distinguishes between three types of samples:
  - ❑ Full-period continuous single samples (integrated).
  - ❑ Full-period consecutive samples (integrated).
  - ❑ Grab samples.

- ❑ The LCL and UCL are calculated differently, depending on the type of sampling method used.

Full-period Continuous Single Sampling

- ❑ Full-period continuous single sampling is defined as sampling over the entire sampling period with only one sample. It could be full-shift or short period ceiling determination.

Full-period Consecutive Samples

- ❑ Full-period consecutive sampling is defined as sampling using multiple consecutive samples of equal or unequal time duration which, if combined, equal the total duration of the sample period (e.g., taking two four-hour silica gel samples).

- ❑ Using multiple samples results in slightly lower sampling and analytical errors.

- ❑ Collection of several samples allows conclusions to be reached concerning the manner in which differing activities performed throughout the workday affect overall exposure.

- If a sample is lost during the sampling period due to pump failure, gross contamination, etc., at least some data will have been collected to evaluate exposure.

Grab Sampling

- Grab sampling is the collection of a number of short-term samples at various times during the work shift which, when combined, provide an estimate of exposure over the total period (e.g., detector tubes, direct reading devices, etc.).

UCL and LCL Method Calculation for Full-period Continuous Single Sample

- A method for calculating the UCL and LCL from full-period continuous single sampling is outlined below.
- Obtain the full-period sampling result (value C_A), the PEL, and the SAE. Divide the sampling result (C_A) by the PEL to determine the standardized concentration (Y).

$$Y = \frac{C_A}{PEL}$$

- Compute the UCL as follows:

$$UCL(95\%) = Y + SAE \quad \textbf{or} \quad UCL = \frac{C_A}{PEL} + SAE$$

- Compute the LCL as follows:

$$LCL(95\%) = Y - SAE \quad \textbf{or} \quad LCL = \frac{C_A}{PEL} - SAE$$

- Classify the exposure according to the following:
  - If the UCL ≤ 1, a violation does not exist.
  - If the LCL ≤ 1 and the UCL > 1, classify as possible overexposure.
  - If LCL > 1, a violation exists.

UCL and LCL Calculation Method for Full-period Consecutive Sampling

- Use the above calculation. If the results fall into the possible overexposure classification, proceed with the more exact formula detailed below.

- Obtain X_1, X_2, X_3... X_n, the n consecutive concentrations on one work shift and their time durations, T_1, T_2, T_3, ... T_n. Obtain the SAE. Compute the TWA exposure. Divide the TWA exposure by the PEL to find Y, the standardized average.

$$Y = \frac{TWA}{PEL}$$

- Compute the UCL(95%) as follows:

$$UCL(95\%) = Y + SAE$$

- Compute the LCL(95%) as follows:

$$LCL(95\%) = Y - SAE$$

- Classify the exposure according to the following classifications:
  - If UCL $\leq$ 1, a violation does not exist.
  - If LCL = 1, and the UCL > 1, classify as possible overexposure.
  - If LCL > 1, a violation exists.

Another Method for Confidence Limits Using Consecutive Sampling

- When sampling has occurred for several distinct time periods, the calculation of the LCL and UCL follows the formula listed below.

$$LCL = C_A - \frac{SAE\sqrt{T_1^2C_1^2 + T_2^2C_2^2 + \ldots T_n^2C_n^2}}{PEL \bullet (T_1 + T_2 + \ldots T_n)}$$

$$UCL = C_A + \frac{SAE\sqrt{T_1^2C_1^2 + T_2^2C_2^2 + \ldots T_n^2C_n^2}}{PEL \bullet (T_1 + T_2 + \ldots T_n)}$$

where:

LCL = lower confidence limit
UCL = upper confidence limit
PEL = permissible exposure limit (mg/m^3 or ppm)
SAE = sampling and analytical error as a decimal
T_n = time period
C_n = concentration during a particular time period
C_A = calculated TWA concentration in air

General Statistical Calculations

Arithmetic Mean

- To average multiple repetitive readings from sampling programs, a simple arithmetic average is used.

$$AM = \frac{X_1 + X_2 + X_3 \ldots + X_n}{n}$$

where:

AM = arithmetic mean
X = value of each observation
n = number of observations

Geometric Mean

- The geometric mean is most common when averaging values that vary over several orders of magnitude.

$$GM = \sqrt[n]{x_1 \bullet x_2 \bullet x_3 \bullet \ldots x_n} \quad \textbf{or}$$

$$GM = 10^{(\log x1 + \log x2 + \log x3 + \ldots \log n)/n} \quad \textbf{or}$$

$$GM = 10^{\Sigma(\log x)/n}$$

where:

GM = geometric mean
x = value of each observation
n = number of observations

Standard Deviation

- The standard deviation expresses the variation from the mean. As discussed earlier, the standard deviation is calculated and multiplied by 100 to achieve the coefficient of variation or CV.

$$SD = \sqrt{\frac{\Sigma(x - x_i)^2}{n - 1}}$$

where:

SD = standard deviation of a sample
x = arithmetic mean
x_i = value of measurement i
n = number of observations

Geometric Standard Deviation

- Particle size distributions of airborne particulates are usually characterized by lognormal distributions rather than simple normal distributions. The geometric standard deviation (GSD) is a dimensionless number outlined in the formula below.

$$GSD = \frac{84.13\ \%\ \text{particle size}}{50\%\ \text{particle size}} \quad \textbf{or} \quad GSD = \frac{50\ \%\ \text{particle size}}{15.87\%\ \text{particle size}}$$

Comparing Means (Student's T-test)

- The comparison of sample and population means the determination of significance can be an elaborate and complicated procedure. When an OSHA compliance officer needs to sample at a site, the site industrial hygienist may choose to sample simultaneously. An industrial hygienist may want to know if the two sets of results are significantly different at the 95% confidence level. Use the formula below when sampling the same process simultaneously with another industrial hygienist.

$$t = \frac{x_1 - x_2}{S_p \sqrt{\frac{1}{n_1} + \frac{1}{n_2}}}$$

$$S_p = \sqrt{\frac{(n_1 - 1)S_1^2 + (n_2 - 1)S_2^2}{n_1 + n_2 - 2}}$$

where:

t	= Student's T-test
x_1	= mean of first set of observations
x_2	= mean of second set of observations
S_1	= standard deviation, set 1
S_2	= standard deviation, set 2
S_p	= pooled standard deviation
n_1	= number of observations, set 1
n_2	= number of observations, set 2

15 Personal Protective Equipment and Respiratory Protection

When it is not feasible to render the work environment entirely free of occupational health hazards, it may be necessary to protect the individual with personal protective equipment. Industrial hygienists always consider the use of personal protective equipment to be a last resort, to be used when engineering and administrative controls are not feasible. Relying on workers to use uncomfortable respirators, hot, bulky clothing, or specialized protective gear is not a desirable method for controlling potential exposures.

Personal Protective Equipment

PERSONAL PROTECTIVE EQUIPMENT (PPE)

- Personal protective equipment (PPE) may be appropriate during short exposures to hazardous contaminants, such as during non-routine maintenance or spill-response procedures.
- The disadvantages of PPE are numerous. These include:
  - Does not eliminate the hazards.
  - Failure results in immediate exposure.
  - May become ineffective without the worker's knowledge.
- OSHA, as a general rule, does not approve different types of personal protective equipment. OSHA does outline regulations for their use, however.
- Employees must receive adequate training to comply with the appropriate usage of PPE.
- In the workplace, the following should be done frequently:
  - Evaluate operations and processes to determine personal protective requirements.
  - Ensure that the PPE available to employees is compliant with recommended standards and is appropriate for the particular hazard.

PROTECTIVE CLOTHING

- A broad range of protective clothing is available: gloves, gauntlets, boots, aprons, coveralls, etc.
- Clothing is available to protect workers from physical hazards (such as ultraviolet light emitted from thermal spray processes) to a wide range of chemical hazards (such as acids, bases, and solvents).

- Body protection clothing ranges from aprons and limited-use disposable coveralls to totally encapsulating chemical protective suits (TECP).
- Chemical protective clothing, whether gloves or totally encapsulating suits, are worn as a barrier to a chemical, physical, or biological hazard that may cause injury if it contacts or is absorbed by the skin.
- No material protects against all chemicals and combinations of chemicals.
- Protective clothing, if not disposable, needs to be decontaminated after use.
- Protective clothing is available in different materials that protect against acids, bases, solvents, oils, etc.
- Standards for personal protection are outlined by the American National Standards Institute (ANSI Z41.1983).

Gloves

- Gloves are the most common form of chemical-protective clothing and should be selected for specific jobs.
- Manufacturers provide a large selection of gloves made of butyl rubber, natural rubber, neoprene, nitrile rubber, polyvinyl alcohol, polyvinyl chloride, Teflon, Viton, and other materials.
- Thick gloves may provide an excellent barrier protection, yet limit mobility and dexterity.
- Impregnated gloves are uniquely made to protect against cuts and abrasions, yet offer no chemical resistance.
- Barrier creams may be used that provide a barrier between the skin and cutting fluids and oils. They may be used as an alternative to gloves that cannot be used around machining operations.

CHEMICAL PROTECTIVE CLOTHING (CPC)

- Chemical protective clothing is manufactured in a wide variety of styles and configurations, ranging from aprons and one-time disposable suits, to totally encapsulating chemical protective suits.
- Chemical protective clothing is worn as a barrier to chemical, physical, or biological hazards that may cause adverse health effects if the material comes in contact with the skin.

Selection of Chemical Protective Clothing

- ❑ When selecting chemical protective clothing, consider the following:
  - ❑ Performance of the protective clothing in exposure reduction.
  - ❑ Physical limitations created by using protective clothing.
  - ❑ Site-specific factors, such as impaired mobility, impaired vision, and heat stress.
  - ❑ Hazards present and the tasks to be performed.
  - ❑ Ease of decontamination (if a material is difficult to decontaminate, consider disposable chemical protective clothing).
  - ❑ Relevant chemical-protective clothing standards.
  - ❑ Cost of chemical protective clothing.
  - ❑ Physical strength of the chemical protective clothing.
  - ❑ Resistance to physical hazards.

- ❑ Establish levels of protection commensurate with the actual or perceived hazards and not based on worst case scenarios that may not pertain to that particular site. Different levels of protection within a work setting make for a more flexible, effective, and less costly operation, while maintaining a high degree of safety.

Effectiveness of Chemical Protective Clothing

- ❑ The effectiveness of chemical protective clothing depends on the clothing's ability to resist permeation, degradation, and penetration.

- ❑ Permeation is the process by which chemicals move through a protective clothing material at the molecular level.

- ❑ Degradation occurs when chemical contact causes deterioration of the physical properties of the protective clothing material, causing swelling or loss of physical strength.

- ❑ Penetration is the direct flow of a chemical through closures, seams, pinholes, or other imperfections in the protective clothing material.

PPE LEVELS

- ❑ Personal protective equipment is categorized into levels A though D. Level A is the most complete and comprehensive level and level D is the lowest level of protection.

Level A Protection

- ❑ Use Level A protection when:
  - ❑ The chemical concentration is known to be above a safe level.
  - ❑ During a confined space entry.
  - ❑ In the presence of extremely hazardous substances (e.g., cyanide).
  - ❑ In the presence of skin destructive substances.

- ❑ Level A protective gear generally includes:
  - ❑ Self-contained breathing apparatus (SCBA).
  - ❑ Hard hat.
  - ❑ Two pairs of gloves.
  - ❑ Chemical resistant steel toe shank and disposable booties.

Level B Protection

- ❑ Use level B protection when:
  - ❑ Immediately dangerous to life and health conditions exist.
  - ❑ Concentrations are above the protection factors provided by a full mask, air-purifying respirator.
  - ❑ Oxygen levels are less than 19.5 percent.
  - ❑ Skin contact is unlikely to the head and neck.
  - ❑ An unidentified vapor is suspected.

- ❑ Level B protective gear generally includes:
  - ❑ Self-contained breathing apparatus.
  - ❑ Hard hat.
  - ❑ Two pairs of gloves.
  - ❑ Chemical resistant steel toe shank and disposable booties.
  - ❑ Two-piece suit with hood or disposable suit.

Level C Protection

- ❑ Use level C protection when:
  - ❑ The air concentration is known.
  - ❑ Assigned protection factors offer control with an air-purifying mask.
  - ❑ There is no threat of immediately dangerous to life and health conditions.
  - ❑ There is no skin destruction.
  - ❑ There is no unidentified vapor present.

- ❑ Level C protective gear generally includes:
  - ❑ Full-face air-purifying mask.
  - ❑ Hard hat.
  - ❑ Two pairs of gloves.
  - ❑ Chemical resistant steel toe with shank and disposable booties.
  - ❑ Two-piece suit or disposable suit.

Level D Protection

- ❑ Use level D protection when:
  - ❑ There is no measurable concentration.
  - ❑ No exposure to splash or inhalation will occur.

- ❑ Level D protective gear generally includes:
  - ❑ Safety glasses.
  - ❑ Hard hat.
  - ❑ One pair of gloves.

- Shoe with steel toe shank.
- Coveralls.

EYE AND FACE PROTECTION

- Eye and face protection includes safety glasses, chemical goggles, and face shields. Appropriate selection is based on the type of hazard.

- Eye and face protection regulations are outlined in OSHA 29 CFR 1910.133. The American National Standards Institute, in its publication *Practice for Occupational and Educational Eye and Face Protection* (ANSI Z87.1989), specifies the use and construction of protective eyewear.

Face Shields

- Face shields or chemical splash goggles are appropriate whenever a worker may be subject to splashing from chemicals. They are excellent for use whenever workers are dipping parts into open-surface tanks, containing plating baths, cleaning solutions, organic chemicals, or corrosive chemicals.

- Face shields are appropriate whenever a worker may be subject to flying particles, such as from using a pedestal grinder.

- Face shields are designed only to prevent direct splash exposures to the face and not to provide complete eye protection.

Safety Glasses

- Safety glasses are necessary when working with hazardous materials or when operating machinery, air guns, or where there is reasonable probability of injury that can be prevented by use of such equipment.

- In most work settings, whenever safety glasses are required, it is implied that those safety glasses include permanently affixed side-shields.

FOOT PROTECTION

- Regulations for foot protection are outlined in the OSHA Foot Protection regulation (29 CFR 1910.136).

- Protective footwear is durable and comfortable footwear designed for expected exposures. Footwear deemed unacceptable in most work settings includes any footwear that permits direct contact between the foot and a foreign agent. Examples include open-toed shoes, open-heel shoes, and soft shoes (e.g., canvas athletic shoes).

- Steel-toed shoes are required whenever there is the potential for items to be dropped onto the foot.

- Rubber boots, boot-covers, or other protective footwear appropriate for chemical splashes are appropriate in areas where hazardous chemicals are used.

HEARING PROTECTION

- Personal hearing protectors, such as earplugs or earmuffs, can provide adequate protection against noise-induced hearing impairment. Hearing protective devices are effective only if properly selected and used.
- Regulations are outlined in the OSHA Occupational Noise Control standard (29 CFR 1910.95). The OSHA standard specifies engineering and administrative controls to be used when feasible. If not feasible, or pending these controls, hearing protection is appropriate.
- Hearing protection users must be trained according to the regulations specified by OSHA. Hearing protection is required whenever employees are exposed to an eight-hour time-weighted-average greater than 85 decibels.
- Hearing protective devices serve as acoustic barriers that reduce the amount of sound energy transmitted through the ear canal to receptors of the inner ear.
- Hearing protection devices generally include earmuffs and inserts. Others include enclosures, entire head protectors and superaural protectors (earcaps).

Inserts

- Inserts are commonly called earplugs. Inserts are inexpensive and have a limited service life. There are usually made of rubber, plastics, fine glass down, foam, etc.
- Types include formable, custom-molded, and premolded.

Earmuffs

- Earmuffs consist of two cup- or dome-shaped devices that fit over the entire external ear, including the lobe, and a cushion or pad that seals against the side of the head. Earmuff efficiency is reduced when the fit is compromised by eyeglass wear.

Disadvantages of Hearing Protection Devices

- Consider the following disadvantages of hearing protection devices when prescribing their use:
  - Hearing protection devices do not reduce or eliminate the source of the noise.
  - Their failure means immediate exposure.
  - They can become ineffective without the knowledge of the user.
  - Training is required on benefits, proper fitting, use, and care.

Hearing Protection Selection

- Considerations for selecting hearing protection include short-term or long-term use (e.g., impact noise vs. continuous noise for an entire shift).
- Each hearing protection device has a listed noise reduction rating (NRR). Manufacturers of the devices provide this information. A manufacturer's reported test data must be examined carefully. They must report the attenuation characteristics of the device over a range of frequencies.
- Appendix B of 29 CFR 1910.95 describes methods of using the NRR to determine whether a particular hearing protective device provides adequate protection within a given exposure environment.

Respiratory Protection

RESPIRATORY PROTECTION OVERVIEW

- Respiratory protection is normally restricted to use in controlling intermittent exposures or for operations where control is not otherwise feasible.
- Respiratory protection offers emergency or short-term protection.
- Respirators are used when necessary to enter a highly contaminated atmosphere for rescue or emergency repair work.
- An approved respirator must be selected for the particular hazard.
- Two general types of respirators are used:
  - Air purifying respirators remove contaminants from breathing air.
  - Air supplied respirators provide clean air from an outside source or from an air tank.
- OSHA requires a respirator program whenever respirators are used in the workplace, as referenced in 29 CFR 1910.134.
- All of the following factors must be considered when selecting the proper type of respirators:
  - Type of air contaminant.
  - Its expected maximum concentration.
  - The possibility of oxygen deficiency.
  - The useful life of the respirator.
  - Escape routes available.

Canister (Air Purifying)

- An air-purifying canister is a container filled with sorbents and catalysts that remove gases and vapors from air drawn through the unit. The canister may also contain an aerosol or particulate filter to remove solid or liquid particles.

Cartridge

- A cartridge is a small container filled with air-purifying media.

Catalyst

- In respirator use, a catalyst is a substance that converts a toxic gas or vapor into a less-toxic gas or vapor.

Filter

- A filter is fibrous media (canned or uncanned) used in respirators to remove solid or liquid particles from the air stream entering the respirator enclosure.

Facepiece

- The facepiece is that portion of a respirator that covers the wearer's nose and mouth in a half-mask facepiece or his or her nose, mouth, and eyes in a full facepiece. It is designed to make a gas-tight or dust-tight fit with the face and includes the headbands, exhalation valve(s), and connections to the air-purifying device or respirable gas source, or both.

Immediately Dangerous to Life and Health (IDLH)

- Immediately dangerous to life and health (IDLH) conditions pose an immediate threat to life or health, such as threat of severe exposure to contaminants or radioactive materials likely to have adverse effects on health.

Resistance

- Resistance is opposition to the flow of air, as through a canister, cartridge, particulate filter, or orifice.

Maximum Permissible Concentration (MPC)

- The National Committee on Radiation Protection sets the maximum permissible concentration (MPC). These MPCs are recommended maximum average concentrations of radionuclides to which a worker may be exposed, assuming that he or she works eight hours a day, five days a week, and 50 weeks a year.

RESPIRATORY PROTECTION PROGRAM ELEMENTS

- The primary objective of workplace respiratory protection is to prevent employee exposure to atmospheric contamination.
- When effective engineering controls are not feasible, or while they are being instituted, appropriate respirators must be used pursuant to the requirements referenced by the OSHA standard, 29 CFR 1910.134.

Employer Responsibility

- OSHA requires that employers provide respirators when necessary. Employers must provide respirators that are applicable and suitable for the purpose intended, and are responsible for establishing and maintaining a respiratory protection program.

Employee Responsibility

- OSHA requires that employees use the provided respiratory protection in accordance with instructions and training received. Employees must guard against damage to the respirator and report any malfunction of the respirator to the responsible person.

Minimal Acceptable Program

- A minimal acceptable program includes written standard operating procedures governing the selection and use of respirators as well as the following:
  - Selection based upon the hazards to which the worker is exposed.
  - Instruction and training in the proper use of respirators, including cleaning and limitations.
  - Regular cleaning and disinfection. Those issued for the exclusive use of one worker should be cleaned after each day's use, or more often if necessary. Those used by more than one worker shall be thoroughly cleaned and disinfected after each use.
  - Storage in a convenient, clean location.
  - Inspections made during cleaning. Worn or deteriorated parts shall be replaced.
  - Respirators for emergency use (such as self-contained devices) shall be thoroughly inspected at least once a month and after each use.
  - Appropriate surveillance of work area conditions and degree of employee exposure or stress shall be maintained.
  - Regular inspection and evaluation to determine the continued effectiveness of the program.
  - Close program administration vested in one individual with sufficient knowledge of the subject to properly supervise the program.

Medical Limitations

- Persons should not be assigned to tasks requiring use of respirators unless it has been determined that they are physically able to perform the work required while using the equipment.

- The local physician shall determine what health and physical conditions are pertinent. Users should be re-evaluated periodically (at least annually).

- Medical evaluation should include spirometry (measuring lung respiratory volumes to determine lung function).

Approved Respirators

- Approved or accepted respirators must be used when they are available.

- Respirators issued must provide adequate respiratory protection against the particular hazard for which they are designed in accordance with standards established by competent authorities.

- Respirators must be tested and listed as satisfactory by an authority having jurisdiction, such as the US Department of the Interior, Bureau of Mines, or the US Department of Agriculture, for respirators manufactured prior to July 10, 1995 and by NIOSH or the Department of Health and Human Services (DHHS) after July 10, 1995.

Classification, Description, and Limitations of Respirators

- Respirators, particularly air-purifying respirators, are designed and selected on the basis of the chemical and physical properties of the air contaminants they protect against.
- Particle size distribution and solubility are important parameters for particulate contaminants.
- Respirators fall into three general classifications, according to mode of operation: air supplying respirators, air-purifying respirators, and combination air supplying/air purifying respirators.
- Air-supplying respirators include self-contained, hose-mask, airline, and combination self-contained and hose-mask or airline.
- Air-purifying respirators include gas and vapor (gas mask and chemical cartridge), particulate (dust, fog, fume, mist, smoke, and sprays), and combination gas, vapor, and particulate types.
- Combination respirators may be both atmosphere-supplying and air-purifying respirators.

RESPIRATOR SELECTION AND USE

- Respirators must be selected on the basis of the hazards to which employees are exposed.
- Use the following considerations when selecting respiratory protection equipment:
  - Identification of the substance or substances for which respiratory protection is necessary.
  - The activities of the workers.
  - The hazards of each substance and its significant physical and chemical properties.
  - The maximum levels of air contamination expected.
  - The probability of oxygen deficiency.
  - The period of time the workers will need to use the respiratory protection devices.
  - The capabilities and physical limitations of the protective device used.
  - The assigned protection factors.

Assigned Protection Factor (APF)

- Each respirator is assigned a protection factor, called the APF. The APF is the minimum anticipated level of protection provided by each type of respirator worn in accordance with an adequate respiratory protection program.

- Example: An APF of 10 means that the respirator should reduce the airborne concentration of a particulate by a factor of 10 (or 10% of the workplace concentration).

- The equation for the factor is outlined below:

$$APF = \frac{\text{Concentration of contaminant outside the mask}}{\text{Concentration of contaminant inside the mask}}$$

Maximum Use Concentration (MUC)

- The maximum use concentration (MUC) is a measure of the degree of protection provided by the respirator to the user. It takes into account the respirator limitation and the ability of the user to get a satisfactory fit.

- Multiplying the assigned protection factor (APF) and the appropriate permissible exposure limit (PEL) will yield the maximum use concentration (APF x PEL = MUC).

- Assigned protection factors include 5 for disposable dust masks, 10 for a half mask air purifying respirator, 50 for a full-face mask when fitted with a quantitative fit test (QNFT) or 10 if fitted with a qualitative fit test (QLFT).

Respirable Air and Oxygen for SCBA and Hose-type Respirators

- Compressed air, compressed oxygen, liquid air, and liquid oxygen used in respirators must be of high purity.

- Oxygen must meet the requirements of the United States Pharmacopoeia for medical or breathing oxygen.

- Breathing air must meet at least the requirements of the specification for Grade D breathing air as described in the Compressed Gas Association Commodity Specification.

- Compressed oxygen that has previously been used as compressed air must not be used. Compressed air might contain low concentrations of oil. When high-pressure oxygen passes through an oil-or grease-coated orifice, an explosion or fire may occur. Oil-less compressors are available. If exposure to oil exists, the compressor must be equipped with an alarm.

- Breathing air may be supplied to respirators from cylinders or air compressors.

- Cylinders must be tested and maintained in accordance with applicable DOT or ICC specifications for shipping containers.

- Compressors must be constructed and situated to avoid entry of contaminated air into the system. Suitable in-line air purifying sorbent beds and filters must be installed to further assure breathing air quality.

- A receiver of sufficient capacity to enable the respirator wearer to escape from a contaminated atmosphere in the event of compressor failure and alarms to indicate compressor failure and overheating must be installed in the systems.

- Airline couplings must be incompatible with outlets for other gas systems to prevent inadvertent servicing of airline respirators with non-respirable gases or oxygen.

- Breathing gas containers must be marked in accordance with appropriate ANSI standards.

Nature of Hazards

- Chemical and physical properties, toxicity, and concentration of the hazardous material must be considered in the selection of respirators.

- To evaluate the nature of the hazard consider:
  - The type of hazard: oxygen deficiency or contaminant (dust, vapor, mist, or gas).
  - Hazardous exposure level: PEL or TLV.

- Is the contaminant flammable? Does the estimated concentration approach the lower explosive limit (LEL) or do dust concentrations create a potential explosion problem?

- Will the contaminant irritate the eyes at the estimated concentration? For irritating materials, a full-facepiece respirator should be employed.

- Can the contaminant be absorbed through the skin as a vapor or liquid?

Oxygen Deficient Atmospheres

- Appropriate respiratory protection equipment for oxygen deficient atmospheres includes:
  - SCBA.
  - Hose mask with blower.
  - Air line respirator with auxiliary self-contained air supply.

- An attendant should stand by at the entrance to the oxygen deficient atmosphere with communications and rescue equipment.

Contaminated Atmospheres

- Contaminated atmospheres require industrial hygienists to know the type of hazard and the atmospheric concentration. Air purifying respirators are usually appropriate for contaminated atmospheres.

- Air-purifying respirators should be available with the appropriate cartridge(s) for dust, vapor, fume, mist, gas, etc.

- If an air-purifying respirator is to be used, the contaminant must have adequate warning properties. The odor threshold must be less than the allowable limit. Example: The odor threshold for MEK is approximately 2 ppm, and the allowable limit is 200 ppm.

Contaminated Atmospheres—IDLH

- Immediately dangerous atmospheres when the contaminant concentration is unknown and/or an air-purifying respirator is not available require more than air purifying respiratory protection. Air supplied respirators must be used. These range from air line respirators, with a remote air supply hosed to the mask, to back-strapped self-contained breathing apparatus (SCBA), such as those used by firefighters.

- Workers must be equipped with safety harnesses and safety lines for lifting or removing persons from hazardous atmospheres or other equivalent provisions for the rescue of persons from hazardous atmospheres must be used. A standby man with SCBA must be available at the nearest fresh air base for rescue.

- Air line respirators are not approved or recommended for use in immediately dangerous atmospheres because no respiratory protection is provided if the air supply fails.

Air Sampling

- Air sampling data is important in the selection of the proper respirator and should include:
  - Identification of the contaminant.
  - Nature of the hazard.
  - Concentration at the breathing zone, including estimation of the TWA.
  - Work time (length of time respirator will be worn (entire shift vs. short periods)).
  - Activity of wearer (heavy activity leads to mouth breathing vs. nasal breathing).
  - Vision (All face pieces can restrict the wearer's vision).
  - Communications.
  - Temperature extremes (sweating can cause silicone half-masks to slip when wearers' sweat).
  - Eye protection.

Respirator Use

- The correct respirator must be specified for each job. This is usually a function of the industrial hygienist. The individual issuing respirators must be adequately instructed on their usage. Date of issuance should be recorded.

Training and Education in Proper Use

- ❑ When training others to use respirators, explain and perform the following:
  - ❑ Nature of the hazard, whether acute, chronic, or both, and an honest appraisal of what may happen if the respirator is not used.
  - ❑ Why more effective control is not immediately feasible.
  - ❑ Recognition that every reasonable effort is being made to reduce or eliminate the need for respirators.
  - ❑ Why this is the proper type of respirator for the particular purpose.
  - ❑ The respirator's capabilities and limitations.
  - ❑ Instruction and training in actual use of the respirator (especially a respirator for emergency use) and close and frequent supervision to assure that it continues to be properly used.
  - ❑ Allow the user to practice wearing the respirator, adjusting the fit, handling it, and testing its facepiece to face seal; wear it for a familiarity period; and finally, to wear it in a test atmosphere.

RESPIRATOR FIT TESTING

- ❑ Respirators should not be worn when conditions prevent a good face seal. Such conditions may be a growth of beard, sideburns, or temple pieces on glasses. The absence of one or both dentures can seriously affect the fit of the face-piece.

- ❑ To assure proper protection, the wearer must check the facepiece fit each time the respirator is worn. Two simple tests include the negative pressure check and positive pressure check.

Qualitative Fit Testing

- ❑ Qualitative fit tests (QLFTs) include subjective methods to determine if a respirator fits the user well. Two examples of quick fit tests that should be used by the wearer each time the respirator is donned include the positive pressure check and the negative pressure check.

- ❑ Other qualitative fit tests, such as the odor test, the irritant smoke test, and the taste test, are used when the respirator is first issued to a new user to simulate realistic conditions of use.

Positive Pressure Check

- ❑ To perform the positive pressure check, close the exhalation valve and exhale gently into the face-piece. Consider the face fit satisfactory if a slight positive pressure can be built up inside the facepiece without any evidence of outward leakage of air at the seal.

Negative Pressure Check

- ❑ To perform the negative pressure check, close off the inlet opening of the canister or cartridge(s) by covering with the palm of the hand(s) or by replacing the seal(s). Inhale gently so that the face-piece collapses slightly, and hold the breath for ten seconds. If the face-piece remains in its slightly collapsed condition and no inward leakage of air is detected, the tightness of the respirator is probably satisfactory.

Odor Test—A Qualitative Fit Test

- A concentration of 100 parts per million isoamyl acetate vapor may be prepared in a special chamber, a small plastic enclosure, or in a vacant room. If the person wearing the respirator can enter and remain in this test atmosphere without detecting the odor of isoamyl acetate, a good fit is indicated.
- Particulate filter respirators can frequently be adapted for use with chemical cartridges and may also be tested for face fit with isoamyl acetate.
- A limitation of the odor test includes the subjectivity of the odor. Some people have different responses to the banana scented oil.

Irritant Smoke Test—A Qualitative Fit Test

- To check the fit of a respirator with a high-efficiency particulate filter, an irritant smoke tube (e.g., a glass tube filled with stannic chloride-impregnated pumice or titanium tetrachloride) produces a very irritating smoke when air is blown through it. The smoke is directed at the face-piece seal and leakage is indicated by irritation of the throat and lungs.
- Irritation of the throat causes the individual to cough and is a good indication of a poor fit.

Taste Test—A Qualitative Fit Test

- The taste test uses saccharin and is conducted in a manner similar to the other two tests. The disadvantage of this test is the wide variation in subjects' ability to taste saccharin, implying different levels of protection for individuals.

Quantitative Fit Testing

- There are several methods of quantitative fit testing available using devices that count specific particles that enter the mask.
- Several regulations require quantitative fit testing, such as the lead standard and the asbestos standard.
- Quantitative fit testing is not subjective and will not cause irritation of the throat like smoke tubes.
- Several products are available for quantitative testing, such as the Portacount.
- Polydisperse sodium chloride aerosol is often used as the challenge agent. Another common challenge is DEHP (Di-ethyl Hexylphthalate) aerosol. DEHP, however, is a suspected carcinogen.
- Advantages include ease of use and speed.

Maintenance and Care

- Respirator users must be informed of the cleaning and maintenance requirements of respirators. These include:
  - Inspection for defects (including a leak check) before and after each use.
  - Cleaning and disinfection.
  - Repair.
  - Where to get spare parts.
  - Storage should be in a clean, dry location, where the respirator cannot be deformed.

Special Problems

- Special problems associated with respirators include:
  - Corrective lenses with full face-piece.
  - Eye wear obstruction with half-mask face-piece.
  - Communications with other workers.
  - Temperature extremes causing fogging and sweating.

Evaluation of Respirator Program Effectiveness

- Feedback from workers on how a respirator program is functioning is necessary.
- The program should be monitored and evaluated on a continuing basis for the following:
  - Wearer acceptance (comfort, ability to breathe without objectionable effort, adequate visibility, provisions for wearing prescription glasses).
  - A qualified individual must conduct examinations of respirators in use (frequent random inspections).
  - Evaluation of protection afforded. (Evaluations of the interiors of the respirators should be made periodically. Solvent vapors can degrade the inhalation valves and the value of the protection.)

PARTICULATE RESPIRATORS

- NIOSH has developed a set of regulations in 42 CFR 84 (also referred to as Part 84) for testing and certifying non-powered air-purifying particulate filter respirators.
- The Part 84 respirators have passed a more demanding certification test than old respirators (e.g., dust and mist (DM), dust, fume, and mist (DFM), spray paint, pesticide, etc.) certified under 30 CFR 11.
- The old Part 11 respirator certification was promulgated in 1972 based upon filter certification tests conducted in the 1930s. New research, testing, and manufacturing technology have made these certification procedures obsolete.

Non-powered Air-purifying Particulate Filter Respirators Only

- The regulatory changes only affect non-powered, air-purifying, particulate filter respirators.

- Certification requirements for all other classes of respirators have been transferred to Part 84 without change. Other respirator testing and certification procedures will be addressed through a series of future changes to Part 84.
- OSHA is allowing the continued use of Part 11 particulate-filter respirators until further notice.

Effective Dates

- The new regulation became effective July 10, 1995, replacing the old Part 11 regulation under which NIOSH and MSHA jointly certified respirators.
- NIOSH allowed selling and shipping Part 11 particulate filters as NIOSH certified until July 10, 1998.

Approval Numbers

- A new sequence of approval numbers (TC-84A-xxxx) is used for non-powered particulate respirators certified under Part 84.
- All other respirator types will continue to use the sequence of approval numbers previously used for Part 11 because the requirements for these other types have not changed.

Certification Labels

- All particulate respirators approved under Part 84 have certification labels bearing NIOSH or Department of Health and Human Services (DHHS) emblems.
- Those approved under Part 11 have either NIOSH or MSHA certification labels.
- This allows the user to distinguish respirators certified before July 10, 1995 under Part 11 from those certified after that date under Part 84.

Three Levels of Filter Efficiencies

- The Part 84 regulation provides three levels of filter efficiency, stated in percentage of particles removed from the air.
- The three levels of filter efficiency are 95%, 99%, and 99.97%.

Three Categories of Resistance

- Each level has three categories of resistance to filter efficiency degradation.
- Filter efficiency degradation is defined as a lowering of filter efficiency or a reduction in the ability of the filter to remove particles as a result of workplace exposure.

- The three categories of resistance to filter efficiency degradation are labeled N, R, and P (N for not resistant to oil, R for resistant to oil, and P for oil proof). The class of filter must be clearly marked on the filter, box, or package.

- Example: A filter marked N95 would mean an N-series filter that is at least 95% efficient.

- 42 CFR 84 eliminates the combination categories of paint spray and pesticide respirator approvals.

Decision Logic Required

- A decision logic matrix is used to select the proper respirator.

- The selection of N-, R-, and P-series filters depends on the presence or absence of oil particles as follows:
  - If no oil particles are present in the work environment, use a filter of any series.
  - If oil particles (e.g., lubricants, cutting fluids, glycerin, etc.) are present, use an R- or P-series filter.
  - If oil particles are present and the filter is to be used for more than one work shift, use only a P-series filter.

- Selection of filter efficiency (i.e., 95%, 99%, or 99.97%) depends on how much filter leakage can be accepted. Higher filter efficiency means lower filter leakage.

- The choice of face-piece depends on the level of protection needed—that is, it depends on the assigned protection factor (APF) needed.

Respirator Selection

- To select the correct respirator for protection against particulates, the following conditions must be known:
  - The identity and concentration of the particulates in the workplace air.
  - The OSHA and MSHA permissible exposure limit (PEL), the NIOSH recommended exposure limit (REL), or other occupational exposure limits for the contaminant.
  - The hazard ratio (HR) (i.e., the airborne particulate concentration divided by the exposure limit).
  - The APF for the class of respirator (the APF should be greater than the HR).
  - The IDLH concentration, including oxygen-deficiency.
  - Any service life information available for combination cartridges or canisters.

- Multiplying the occupational exposure limit by the APF for a respirator gives the maximum use concentration in which that respirator can be used.

Use Limitations

- Service life of filters in all three categories of filter efficiency degradation is limited by considerations of hygiene, damage, and breathing resistance.

Filter Certification

- The NIOSH filter certification test is called worst-case (i.e., it produces maximum filter penetration) because the test conditions are the most severe that are likely to be encountered in a work environment.

- These conditions are:
  - Air flow that simulates a high work rate (85 ± 4 lpm for single filters, 42.5 ± 2 lpm through each filter for paired filters).
  - The most penetrating aerosol size (approximately 0.3 micrometer).
  - Charge-neutralized particles.
  - The most filter-degrading test aerosol for R- and P-series filters.
  - Measurement of instantaneous (not average) penetration.
  - High total filter loading (up to 200 mg for N- and R-series filters, and continued loading until there is no further decrease in efficiency for P filters).

- The challenge agents for testing will either be sodium chloride (NaCl) or dioctyl phthalate (DOP) as the test aerosol.

Protection for TB

- The only respirators certified by NIOSH under Part 11 that meet CDC filtration efficiency performance criteria for protection against tuberculosis (TB) are those with HEPA filters. All nine classes of non-powered, air purifying, particulate-filter respirators certified under Part 84 meet or exceed the CDC filtration efficiency performance criteria.

16 Industrial Processes and Associated Health Hazards

Industrial hygienists must understand industrial processes and their impact on workers and the workplace environment in order to be able to identify potential health hazards. Consideration must be given to the materials used, processes, exposure time, and controls.

Abrasive Blasting

ABRASIVE BLASTING AND ITS APPLICATIONS

- Abrasive blasting is used in building construction, shipbuilding, founding, and metal finishing.
- It can be used to remove surface coatings, scale, rust, flashing, tooling marks, and welded or machined fabrications.
- Steel shot is used to peen (hammer) metal surfaces, improving their resistance to fatigue stress, reduce surface porosity, and increase wearability of parts.

Abrasives Used

- Common abrasives include glass beads, silica sand, plastic, nut shells, ground corn cobs, aluminum oxide (Al_2O_3), silicon carbide (SiC), and iron and steel shot.

Blasting Cabinets

- A variety of exhaust enclosures are used in the metal working industry including cabinets, blasting rooms, rotating tables, and tunnel units. These units integrate local exhaust ventilation systems.
- The wet blast technique uses a suspension of an abrasive in water or a petroleum distillate directed at the work piece from a nozzle, using water pressure or air and water pressure.

Open Air Blasting

- Air blasting uses an abrasive in a pressurized air tank that is fed to the blasting hose. A nozzle directs the mixture to the work piece at high velocity.
- Open air blasting is used in shipyards and construction. General area contamination occurs unless the work area is enclosed with a temporary barrier. Blasting crews, operators (pot men), and clean up personnel may be exposed to high dust concentrations during the course of their work.

- Open air blasting is difficult to control, and may cause silicosis in blasters and nearby workers.

- In some cases, the blasting equipment contains a low volume/high velocity exhaust system attached to the abrasive nozzle that collects spent abrasive and work piece debris.

Health Hazards

- Inhalation of airborne contaminants depends on the:
  - Base metal being blasted (usually iron or steel, resulting in exposure to iron dust and other metal alloys including nickel, manganese, lead, and chromium).
  - Nature of the surface being removed (surface could be a lead-based paint or fused silica sand).
  - Abrasive in use and its contamination from previous use (abrasives get old and surface material may contaminate blasting material, e.g., lead paint chips in sand).

- Noise can range from 85 to 110 decibels from blasting cabinets. Open air blasting can range from 85 to 118 decibels.

Abrasive Blasting Engineering Controls

- Engineering controls for abrasive blasting include local exhaust ventilation, rugged enclosures, easily maintained curtains, and seals eliminating outward dust leaks.

- Dust control depends upon the integrity of the enclosure. These enclosures need periodic inspections of seals, doors, windows, gloves, and sleeves.

Abrasive Blasting PPE Controls

- Hearing protection is a necessary personal precaution for abrasive blasting operations.

- Obtain NIOSH approved type CE abrasive blasting hoods for blasting rooms. Air intake to an air line respirator must be located in a contaminant free area.

- Air-purifying respirators with HEPA filters may be required for blasting cabinets during use or clean up, depending on air concentrations.

Acid and Alkali Cleaning of Metals

ACID PICKLING AND BRIGHT DIP

- Acid pickling and bright dips condition parts and products for electroplating or other finishes.

- Pickling is the treatment of metals with acids. It is named for the early process of dipping metals in vinegar. Pickling removes scale, rust, and corrosion products.

- Scale and rust are removed from low and medium alloy steels using a nonelectrolytic immersion bath of:
  - 5-15% sulfuric acid at 140-180° F.
  - 10-25% hydrochloric acid at room temperature.
  - Phosphoric acid to remove hard scale and water scale.

- Pickling film is removed with an anodic electropickle, alkaline electrocleaning, or ultrasonic cleaning.

Variety of Pickling Processes

- The pickling process varies depending upon the type of metal cleaned. Efficiency can be improved ultrasonically, electrolytically, or by operating at higher temperatures. Following pickling, a passivation process is conducted by placing metals in nitric acid to provide corrosion resistance.

- Stainless steels that have been heat-treated need to have the oxide film that forms removed with a two-stage process. Example: Stainless steel can be cleaned in a solution of chromium and heated sulfuric acid to soften scale, then bathed in a mixture of hydrochloric, nitric, and hydrofluoric acids at room temperature.

- Excessive loss of metal or pitting may occur in pickling operations. Several inhibitors are used to reduce the attack.

Bright Dips

- Acid bright dips are usually mixtures of nitric and sulfuric acids used to provide a mirror-like surface to cadmium, magnesium, copper, copper alloys, silver, and sometimes stainless steel.

- Nitrogen oxide gases are usually emitted from these baths.

Emissions from Pickling Operations

- Various acid mists and gases are released from pickling operations and bright dips, depending on:
  - Bath temperature.
  - Surface area of work.
  - Current density (if bath is electrolytic).
  - Whether bath contains inhibitors that produce a foam blanket on the bath or lower the surface tension of the bath, thereby reducing misting.

Controls for Pickling Hazards

- Pickling baths should be ventilated with local exhaust ventilation. Due to carryover of pickling acids in the rinse, the rinse tanks may need local exhaust ventilation. Bright dip baths require more efficient hooding and exhaust than do most other acid treatment operations.

- Administrative controls include:
  - Prohibiting eating, drinking, and smoking in the work area.
  - Authorizing only trained employees to make additions to the baths.
  - Requiring face shields, goggles, gloves, aprons and plater's boots to be worn when adding chemicals to baths. A respirator should also be worn, depending on air sampling results.

Salt Baths

- A bath of molten caustic at 700-1000° F is used for initial cleaning and descaling of cast iron, alloy, copper, aluminum, and nickel. The bath is followed by quenching and acid pickling.

- Metal oxides are removed without attacking the base metal. Oils are burned off. Debris and sludge fall to the bottom; and contaminants float to the surface and are volatilized as vapor or fume.

- All molten baths require subsequent quenching, pickling, and rinsing. The quenching operation dislodges the scale through steam generation and thermal shock.

- Local exhaust is necessary and the tank must be equipped with a complete enclosure to protect the operator from violent splashing.

ALKALINE TREATMENT

- Alkaline immersion systems are used in soak, spray, and electrolytic cleaning used for removal of oil, grease, and paint. These solutions include sodium hydroxide, potassium hydroxide, sodium carbonate, sodium meta or ortho-silicate, trisodium phosphate, and tetrasodium pyrophosphate.

- The electrolytic alkaline cleaning process usually follows a primary cleaning. The initial soak is used to loosen the soil. As in an electrolytic cell, the water dissociates and oxygen is released at the anode and hydrogen at the cathode. The process is powered with direct current, with the work piece conventionally the cathode and an inert electrode as the anode; when the system is powered, the hydrogen gas generated at the work piece causes agitation and removes surface grime.

- As in electroplating baths, gases generated at the electrodes may result in the release of sodium hydroxide mist and steam at the surface of the bath. Mist generation is greatest under cathodic cleaning (if the work piece is the cathode) and varies with bath concentration, temperature, and current density.

- Surfactants and additives that provide a foam blanket on the bath are important to minimize gaseous releases from the bath. Small plastic "ping pong"-type balls have demonstrated some effectiveness in mist control.

Degreasing

DEGREASING AND ITS APPLICATIONS

- Cold degreasing, or soak cleaning, uses solvents to remove surface grime, oil, and grease from metal.

Cold Degreasing

- Cold degreasing is used routinely for the cleaning of metal parts.

- Solvents vary from high flash point petroleum distillates to mixtures that include aliphatic and aromatic hydrocarbons, chlorinated hydrocarbons, ketones, cellosolves, and citrus-based products.

- No universal controls can be established, except that no readily volatile or fast-drying solvent should be routinely used without effective mechanical exhaust ventilation. Avoid skin contact and protect workers from splashing.

Power Flushing

- Spraying with high flash point petroleum distillates such as Stoddard solvent, mineral spirits, or kerosene are widely used to degrease oils from metals.

- Solvents used for this purpose should not have flash points less than 100° F. If this technique is used, a small booth with suitable local exhaust ventilation is preferred, which may or may not be fitted with a fire door and automatic sprinklers.

Electroplating

ELECTROPLATING AND ITS APPLICATIONS

- Metal, plastic or rubber parts are electroplated to prevent rusting and corrosion, to improve appearance, to reduce electrical contact resistance, to provide electrical insulation, and to improve wearability.

- Prior to plating, parts must be cleaned as described above. Common plating metals include cadmium, chromium, copper, gold, nickel, and silver.

Electroplating Technique

- The plating tank contains an electrolyte consisting of a metal salt of the metal to be deposited dissolved in water.
- Two electrodes powered by a low-voltage dc power supply are immersed in the electrolyte. The cathode is the work piece to be plated and the anode is usually a bar of the metal to be deposited. When power is applied, metal ions deposit out of the bath on the cathode or the work piece.
- Water is dissociated, releasing hydrogen at the cathode and oxygen at the anode. The anode replenishes the metallic ion concentration in the bath. Current density (expressed in amperes per unit area of work piece surface) varies depending on the operation.
- Individual parts or racks of small parts are hung manually from the cathode bar. Parts are usually transferred from tank to tank manually. Higher production shops may use automation.

Anodizing Technique

- Anodizing is a surface treatment for decoration, corrosion resistance, and electrical insulation on such metals as magnesium, aluminum, and titanium.
- The work piece is the anode and the cathode is a lead bar. The oxygen formed at the work piece causes surface oxidation.
- A chromic acid bath with high current density serves as the process medium. Since its efficiency is low, the amount of misting is high.

Air Contaminants

- Air contaminants in electroplating result from the release of the bath electrolyte to the air by the gassing of the bath. The gases bubbles released at the electrode rise to the surface of the bath and break, generating a fine airborne mist.
- The rate of mist generation depends upon bath efficiency. For examples:
  - Copper plate is nearly 100% efficient and releases little mist.
  - Nickel plating operates at 95% efficiency and gassing is minimal.
  - Chromium plating is quite inefficient and up to 90% of the total energy is dissociated, with resulting severe gassing and potential exposure to chromic acid mist.

- Although contaminant generation is governed by bath efficiency, it also depends on:
  - Metallic ion concentration.
  - Current density.
  - Nature of bath additives.
  - Bath temperature.
  - Air or mechanical agitation of the bath used to improve plating quality.

Health Significance

- The health significance of electroplating depends on bath contents.
- An inventory of the chemicals in common electroplating baths, the form in which they are released to the air, and the rate of gassing is included in the ACGIH *Industrial Ventilation Manual.*

General Controls

- Several commonly used controls for electroplating processes include mist reduction, personal protective equipment, and proper handling methods.
- Mist reduction can be achieved through:
  - Proprietary bath additives that are available to reduce the surface tension of the electrolyte and reduce misting. Thick foam traps the mist released from the bath.
  - A layer of plastic chips, beads, or balls on the surface of the bath also traps the mist and permits it to drain back into the bath.
- PPE and handling:
  - Major safety hazards occur during handling of concentrated acid and alkali chemicals for tank replenishment.
  - Extensive personal protective equipment is required for electroplaters. Hazards include contact with skin and dermatitis (nickel itch).

Local Exhaust Ventilation (LEV) Controls

- Local exhaust ventilation is necessary for electroplating operations. These systems are frequently end takeoff, lateral slot, or upward plenum slotted hoods.
- Ventilation requirements are determined using a minimum control velocity based on the hazard potential of the bath and the rate of contaminant generation.
- Exhaust volume is calculated using this control velocity and the tank's measurements and geometry.

- Ventilation evaluations using pitot tubes are necessary. Observed values are then compared to recommended values. It is possible to measure the hood slot velocity, average the readings, and calculate the exhaust volume from these measurements. This technique has limited value due to the difficulty in defining average velocity in a narrow slot.

- Measurement of slot velocity is useful in determining the uniformity of the exhaust over the length of the tank.

- Ventilation systems need to be periodically examined for corrosion.

- Adequate makeup air is required as is suitable air cleaning for chromic acid and alkali mists releases.

LEV Exceptions

- Not all processes require local exhaust ventilation. In some facilities, hot water rinse tanks are not ventilated. In other facilities, all heated tanks are ventilated.

- Neither sulfate nor chloride plating baths for nickel require exhaust ventilation, since gassing is negligible.

- Copper plating in an acid solution is conducted with low current density and no agitation. LEV may not be required under such circumstances. But when high current densities with agitation are used, LEV is required.

- Neither zinc nor cadmium plating solutions require LEV unless high current densities and high bath temperatures exist.

Grinding, Polishing, and Buffing

GRINDING, POLISHING, BUFFING, AND THEIR APPLICATIONS

- Grinding techniques include:
  - Cutoff operations in foundries where gates, and risers are removed.
  - Rough grindings of forgings and castings.
  - Facing off weld markings.
  - Grinding out major surface imperfections in metal finishing.

- Common abrasives include aluminum oxide, silicon carbide, resinoid bonding wheels, and rubber and shellac combinations.

- Bonding agents include vitrified glass, sodium silicate, rubber, and shellac. The abrasive is bonded to the surface of a belt, disk, or wheel.

- Grinding wheel composition must be known to determine the possible air contaminants released. Wheel components normally make up a small fraction of the total airborne particulates released during grinding. The bulk of the particulates are released from the work piece itself.

- Grinding is the first phase of the grinding, polishing, and buffing process. Polishing removes surface imperfections. Buffing removes the least amount of material from the work piece and provides a high luster surface.

Potential Exposures in Grinding Operations

- Exposures to grinding dust can be evaluated by means of personal air sampling with gravimetric analysis.

- Exposures include:
  - Toxic metals such as antimony, beryllium, chromium, cobalt, lead, tungsten, and vanadium.
  - Metals coated with lead or chromate-based paints.

Control of Grinding Hazards

- The principal dust control is ventilation. Hood designs utilize a tight hood enclosure with minimum wheel-hood clearance to provide dust control at a minimum exhaust volume.

- Bursting wheel hazards (during high speed operations) and fire hazards of certain metals (magnesium and titanium) affect the design of hoods and wet dust collection systems.

Heat Treating

HEAT TREATING AND ITS APPLICATIONS

- A range of metal alloy heat treating methods are available to improve the strength, impact resistance, hardness, durability, and heat and corrosion resistance of the work piece.

- The most common hardening procedure is heating the work piece to a high temperature, followed by rapid cooling. Softening is accomplished by heating only or heating followed by slow cooling.

Hardening

- Case hardening, the production of a hard surface or case to the work piece, is achieved by diffusing carbon or nitrogen into the metal surface to a given depth.

- This process may be accomplished in air, in furnaces, or in immersion baths by carburizing, cyaniding, or gas nitriding.

Carburizing

- The work piece is heated to 1600-1800°F in an atmosphere containing high concentrations of carbon monoxide (CO), which is the source of diffused carbon.
- In gas carburizing, the parts are heated in a furnace with hydrocarbon gases or CO.
- The principal hazard in carburizing is exposure to carbon monoxide and products of combustion.
- Since gas is supplied to the furnace, small leaks may result in significant workroom exposure.
- If parts are not clean, volatile material will be driven off in the furnace and will generate toxic air contaminants that may leak into the workplace.

Cyaniding

- Cyaniding or liquid carbon nitriding is metal immersion in a cyanide bath with a subsequent quench.
- The metal part is commonly held in a sodium cyanide bath at temperatures above 1600° F for 30 to 60 minutes.
- Air contaminants released include sodium carbonate. Cyanide compounds are not released.
- Handling cyanide salts requires dry storage, separation from acids, and planned disposal.

Gas Nitriding

- Gas nitriding diffuses liquid nitrogen in metal. The furnace atmosphere of ammonia operates at 900-1050° F.
- The handling of ammonia may result in fire, explosion, and toxicity hazards.

Annealing (Tempering Baths)

- Tempering or salt baths may be composed of nitrate salts or a mixture of nitrate and nitrite salts operating in the 400-900° F range. This is used for aluminum and special steel alloys.

Quenching

- Hardening of metals requires a rapid quench. The quench baths may be water, oil, molten salt, liquid air, or brine.
- Potential problems range from nuisance steam to acrolein or other thermal degradation products from oil.

- Local exhaust ventilation may be necessary.

Industrial Radiography

INDUSTRIAL RADIOGRAPHY AND ITS APPLICATIONS

- Radiography is used in industry for the examination of metal fabrications, such as weldments, castings, and forgings in a variety of settings.
- It is used to locate inconsistencies in the object or part.
- Specially designed shielded cabinets are usually placed within manufacturing areas for the in-process examination of parts.
- Radiography may also be conducted at construction sites with more portable equipment.
- Radiography exposes an object to be examined to x-rays or gamma rays from one side and measures the amount of radiation that emerges from the opposite side.
- As radiation passes through the examined part, some of the radiation is absorbed or attenuated. The degree of attenuation depends on the thickness of the part, its density, and its atomic number.

Gamma Rays

- Gamma rays used in industrial radiography are produced as a result of the decay of radioactive nuclei.
- Principal radioisotopes are iridium-192 and cobalt-60. Radioisotope sources produce gamma rays with discrete energies, in contrast with the continuous spectrum of energies produced by x-ray generators.
- Due to radioactive decay, the activity of radioisotope sources decreases exponentially with time. These sources do not require any external energy, which makes them attractive in remote locations.
- The radioactive isotope is sealed inside a source capsule (usually stainless steel). The source capsule is stored in a shielded container, or “pig,” when not in use.

X-rays

- These devices use high voltage power supplies for the production of x-rays, and consideration must be given to electrical hazards.
- OSHA and many states regulate the use of industrial x-ray generators.

- Areas in which radiography is performed must be posted with signs that bear the radiation caution symbol and a warning statement. Access to these areas must be secured for authorized entry only.

- Operators are required to wear personnel monitoring devices to measure the magnitude of their exposure to radiation. Typical devices are film badges, thermoluminescent dosimeters, and direct reading pocket dosimeters. Pregnant workers must notify their employer of their condition and wear a dosimeter commonly referred to as a "belly badge."

- Operators must be trained in the use of radiation survey instruments to monitor radiation levels.

Use of Radioactive Isotopes

- The Nuclear Regulatory Commission (NRC) or the individual state regulatory commission must license users of radioactive isotopes.

- Operators must:
  - Receive training.
  - Wear a film badge or thermoluminescent dosimeter, and a direct reading pocket dosimeter.
  - Use a calibrated radiation survey instrument during all radiographic operations.

- Work areas must be posted with radiation warning signs, and access to these areas must be secured as described earlier.

Metal Machining

TOOL MACHINING

- The fabrication of metal parts from solid stock is done with a variety of machine tools, including the lathe, drill press, miller, shaper, planer, and surface grinder.

- Machining operations involving turning, milling, and drilling use cutting tools that shave off metal from the work piece, forming a thin running coil that normally breaks at the tool to form small chips.

- Temperature and pressure extremes also occur at the interface between the cutting tool and the work.

- To cool this cutting point, a lubricant flushes away the chips formed by the cutting tool. The lubricant or cutting fluid is directed to the cutting point in a solid stream (flood) or a mist to cool the contact point.

Airborne Particulates Generated

- Airborne particulates generated by machining operations depend on the following:
  - Type of base metal and cutting tool.
  - The dust- forming characteristics of the metal.
  - The machining technique.
  - The coolant, and its application method.
- Commonly-machined metals range from mild steel (with no potential health hazards) to various high temperature and stainless alloys (containing toxic metals, such as chromium, nickel, and cobalt).
- Highly toxic metals (such as beryllium) present significant exposures that require vigorous control in any machining operation.
- The metal particulate airborne concentrations from conventional metals and alloys are usually minimal.

The Cutting Tool

- A range of specialized alloys are used in cutting tools. These include:
  - High carbon steels with alloying elements of vanadium, chromium, and manganese.
  - High speed steels containing manganese and tungsten.
  - Special cobalt steels.
  - Cast alloys of tungsten, chromium, and cobalt.
  - Tungsten carbide.
  - Ceramics.
  - Superalloys containing high amounts of nickel.
- Material loss from the cutting tool in conventional machining does not represent a potential hazard.
- Local exhaust ventilation should be provided when preparing cutting tools for use.

The Coolant

- Coolants and cutting fluids cool and lubricate the point of contact and flush away chips. These fluids are presently available as:
  - Soluble (emulsified) cutting oils based on mineral oil emulsified in water with soaps or sulfonates.
  - Straight cutting oils based on mineral oils, with the addition of fatty acids and severely hydro-treated oils.
  - Synthetic oils of varying composition.

Skin Hazards

- Cutting oils may cause acne lesions, infection of sweat pores, and hair folliculitis. Occasionally, skin sensitization occurs.
- Bacterial contamination of cutting fluids does not cause dermatitis. However, it can exacerbate skin irritation.
- The fluids may become contaminated with trace quantities of the base metal that is being worked on. The worker then contacts these dissolved metals in the fluid. Chromium, nickel, and cobalt are strong sensitizers.
- Skin cancer has been linked to machining operations using mineral oil lubrication.
- Operators must keep coolants clean, measure their concentration (for synthetic and semisynthetic types), and must not spit chewing tobacco or other debris into the coolant recirculating units.

Inhalation Hazards

- The contact between cutting fluid and hot, rotating parts generates an oil mist, causing the characteristic smell of machine shops.
- Cutting fluids containing both nitrates and amines could result in the generation of various carcinogenic nitrosamines. Controls involve:
  - Using coolants that do not contain nitrates.
  - Maintaining oil mist concentrations below the 5 mg/m^3 standard.
  - Using anti-mist additives.
  - Providing hood enclosures with suitable exhaust ventilation and air cleaning.

Fire Hazards

- Machining operations involving magnesium and titanium may generate explosive concentrations of dust. These operations must be conducted with suitable ventilation and air cleaning.

ELECTRO-CHEMICAL MACHINING

- The electrochemical machining (ECM) process is similar to the electroplating process.
- In the ECM process, the work piece is the anode, and the cathode is machined to reflect the geometry of the hole to be cut in the work piece.
- ECM is fast, does not produce burrs, and results in little tool wear. It is an excellent process for cutting irregularly shaped holes in hard metals.
- Local exhaust ventilation is necessary to ensure hydrogen concentrations do not approach the lower flammability limit.

Electrical Discharge Machining

- A spark-gap technique is the basis for the electrical discharge machining process (EDM).
- This technique is used for drilling small holes in complex parts.
- The work piece (anode) and the tool (cathode) are immersed in a dielectric oil bath. A voltage across the gap increases until breakdown occurs and there is a spark discharge across the gap that produces a high temperature at the discharge point. This spark, applied repeatedly at 200 to 500 hertz, erodes a small quantity of the metal from the work piece.

EDM Hazards

- EDM hazards are minimal and are usually associated with the oil. A petroleum distillate such as Stoddard solvent or a mineral oil serves as the dielectric.
- When heavy oil mists are present, local exhaust ventilation is needed. The oil gradually becomes contaminated with metal eroded from the part. These metals may dissolve in the oil and create a dermatitis problem.

Metal Thermal Spraying

METAL AND OTHER SPRAYED MATERIALS

- A fine mesh powder or metal wire is melted in an application gun and the molten particles are sprayed to a work piece at high velocity. There they freeze and adhere to the work piece surface. Metal sprays are used for:
  - Corrosion protection.
  - Building up worn or corroded parts.
  - Improving wear resistance.
  - Reducing production costs.
  - Making decorative surfaces.
- Ceramics have largely replaced metals in industries (such as aerospace) where coated parts must experience extremely high temperatures.
- Sprayed materials include:
  - Metals: tungsten, nickel, chromium, tantalum, aluminum, zirconium, iridium, cadmium, lead, molybdenum, and yttrium.
  - Ceramics: alumina, zirconia, beryllia, spinel, zircon, glass, and yttria.
  - Alloys: nichrome, stainless steel, babbitt, and bronze.
  - Intermetallic compounds: nickel aluminide and molybdenum disulfide.
  - Interstitial compounds: titanium carbide and chromium carbide.

Types of Thermal Applications

- There are several types of thermal spraying techniques. These include combustion spraying, detonation spraying, and plasma spraying.

Thermal Spray Hazards

- The major hazard common to all techniques of spraying is the potential exposure of workers to toxic metal fumes.
- If the vapor pressure of the metal at the application temperature is high, resulting air concentrations of the metal are high.
- The percentage of the metal spray deposited on the work piece varies with application techniques and coating metals.
- Overspray varies with the techniques and metals used. Plasma arc spray with chromium carbide exceeds 50% overspray, yet alumina-titania overspray is less than 10%.
- Air samples should be taken to determine worker exposure.
- Other potential hazards include exposure to toxic gases, infrared and ultraviolet radiation, and noise. Plasma spray presents the most hazards. Combustion spraying has the least hazards.
- Appreciable noise reduction can be achieved by changing the operating characteristics of the spray equipment. Lowering the gas flow rate on wire and powder guns can reduce decibels by 3 to 5dB.
- Robotic metal spray equipment that effectively separates the worker from the spraying process is available.
- Local exhaust ventilation is necessary. Respirators may be necessary for manual spraying of toxic metals. Skin protection from ultraviolet radiation may also be necessary.

Nondestructive Testing

NON-DESTRUCTIVE TESTING (NDT) AND ITS APPLICATIONS

- Non-destructive techniques (NDT) include dye penetrants and magnetic particle inspection.
- NDT is commonly used in industry without significant health hazards. But cleaning and preparation of materials for inspection by these methods may present health hazards.

Penetrant Inspection

- NDT is used to inspect surface imperfections in any nonmagnetic solid. It is used to inspect incoming material, such as bar stock and forgings, and for in-process final inspection.

- Fluorescent or visible dyes are suspended in a liquid carrier that is applied to the surface of the part by brushing, dipping, or spraying.

- At surface imperfections, the penetrant is drawn into the imperfection. The part is rinsed and a dry absorbent powder is applied by brushing or dipping.

- The penetrant in the imperfection seeps into the powder and is identified clearly by direct viewing if a visible dye is used or with an UV lamp in a dark room in the case of the fluorescent dye.

- Diverse materials are used as dyes and carriers.

Magnetic Particle Inspection

- Magnetic particle inspection complements the fluorescent penetrant system. The procedure is used for incoming, in process, and final inspection.

- Ferromagnetic particles are applied to the surface of a magnetic material by an air powder gun or by dipping in a bath containing the particles suspended in light petroleum oil or water.

- In magnetic fields induced by a low voltage, high current power supply, the particles are attracted to any surface discontinuity and deposit on its edges, thereby defining geometry.

- The magnetic particles, designed to fluoresce, can be identified with UV radiation, usually from a low energy lamp.

- The health hazards are minimal. Dipping rather than spray application is encouraged.

Painting

PAINT AND ITS APPLICATIONS

- Paint products are used widely to provide a surface coating for protection against corrosion, to enhance appearance, and as electrical insulation.

- Paint commonly refers to a range of products, including conventional paints, varnishes, enamels, and lacquers.

Paint Composition

- Solvent-based paints consist of the vehicle, filler, and additives. The vehicle represents the total liquid content of the paint and includes the binder (to form film) and the solvent (as a thinner).

- The filler includes:
  - Pigments to add color and build body (e.g., white pigments include white lead and zinc oxide, green pigments include chrome green and chromium oxide).
  - Extenders to build body (e.g., talc, mica, china clay).

- The additives include:
  - Dryers to speed drying or curing.
  - Biocides to prevent growth of mold.
  - Flatting agents to reduce luster.
  - Stabilizers to protect against heat and UV.
  - Antiskinning to prevent skin formation in can.

- The heavy metal compounds used as pigments in fillers have historically presented health hazards. Lead, chromates, cadmium, and sulfates have been curtailed, but they are still present in certain paints (especially high temperature paints). Zinc phosphate has replaced lead.

Epoxy Paint Systems

- Two-component urethane paint systems consist of a polyurethane pre-polymer (containing a reactive isocyanate) and polyester. Mixing the two materials initiates a reaction, forming a chemically resistant coating.

- Single package systems use a blocked isocyanate. Unblocking occurs during high temperature baking, when the isocyanate is released to react with the polyester. These paints remain stable at high temperatures, resist abrasion, and have excellent adhesive properties.

- The conventional epoxy system is a two-component system consisting of a resin (based on the reaction products of bisphenol A) and epichlorohydrin. The second component is the hardener or curing agent.

- Epoxy systems are known for their skin sensitivity properties.

Painting Operations

- Paints are applied to parts by brush, roller, dip, flow, tumbling, conventional air spray, airless spray, atomization, and powder coating.

- Conventional air spraying is the most common method and presents the principal hazards, due to overspray and rebound.

- Airless electrostatic spraying and hot spray techniques minimize mist and solvent exposure to the operator.

- The electrostatic technique places a charge on the paint mist particle so that it is attracted to the part to be painted. (This technique is used on appliances, file cabinets, etc.)

- New success with powder coating results in less exposure to operators and minimizes the generation of hazardous wastes.

Potential Exposures

- Exposure to solvent and thinner occurs during brushing, in dipping, and during drying of the parts.

- During spraying, the worker is exposed to the solvent and paint mist.

Controls for Painting Operations

- Local exhaust ventilation is required in most industrial paint application processes. Drying stations and curing ovens require local exhaust ventilation. The usual ventilation control for spray application is a spray booth, room, or tunnel with some type of mist arrestor.

- Booths are commonly equipped with a water curtain or a throwaway dry filter to provide paint mist removal.

- Skin contact must be minimized to prevent defatting of the skin from solvents and primary irritation, as well as skin sensitization from two-component polyurethane systems.

Soldering and Brazing

SOLDERING

- Soldering is the joining of metals using a filler metal.
  - Soft solder uses filler metal with a melting point less than 600° F.
  - Hard solder uses filler metal with a melting point between 600-800° F.
  - Brazing uses filler metal with a melting point greater than 800° F.

- The process includes cleaning the base metal and other components for soldering (fluxing), the actual soldering process, and postsoldering cleaning.

Soldering Flux

- All metals have a film of tarnish that must be removed to effectively wet the metal with solder and accomplish a good mechanical bond.

- The tarnish consists of corrosion, oxidation, carbonates and sulfides.

- The flux removes any adsorbed gases and tarnish from the surface of the base metal and keeps it clean until the solder is applied.

- The flux is usually a corrosive cleaner in a solvent. Rosin is a common base for organic flux. Common flux materials are inorganic (acids, salts), organic non-rosin (acids, halogens, amines), and organic rosin base fluxes (rosin or resin with an activator).

- The flux is applied by brush, roll, spray, foam, dip, paste, or cored solder. Spraying techniques may require local exhaust ventilation.

Solder

- The most common solder contains 65% tin and 35% lead.

- The melting point of solder is low. At these temperatures, the vapor pressure of lead usually does not result in significant air concentrations of metal fumes.

Health Hazards

- Potential health hazards are minimal. Simple tin-lead solder is used routinely at temperatures considered too low to generate significant concentrations of lead fume.

- Activated rosin fluxing agents may require local exhaust ventilation if thermal degradation products are released. Handling solder dross during cleanup may result in exposure to lead dust. Flux represents the most significant potential hazard from soldering.

BRAZING

- Brazing is sometimes referred to as soldering. It is more properly identified as a welding process.

- Brazing is commonly used in appliance manufacture, aerospace component manufacture, and jewelry making.

- Brazing joins similar and dissimilar metals. The brazed joint looks like a soft solder bond but it is a much stronger bond. The temperature of the operation determines the vapor pressure of the metals that are heated and the concentration of metal fumes to which an operator is exposed.

Flux

- The flux is chosen to prevent oxidation and not, as in soldering, to prepare the surface.

- Common fluxes are fluorine, chlorine, and phosphorous compounds. They present the same hazards as soldering fluxes. Flux is corrosive to skin and can cause respiratory irritation.

Application

- Small jobs can be done with a torch. Large production jobs require a brazing furnace or induction heater. Brazing temperatures define the hazard of the various operations. Filler metals with higher brazing temperatures will present the most severe health hazards.
- Brazing furnaces are used widely for the high production parts. Parts are positioned in a holding jig and flux and filler pre-forms are positioned at the brazing locations.
- There is no significant health hazard associated with furnace operations.
- Furnace insulation may contain refractory ceramic fibers. Brazing ovens may be lined with refractive ceramic brick that decomposes over time at high temperatures and must be periodically replaced. These fibers, when heated above 2100° F repeatedly, degrade to free silica. Over time, oven insulation may become friable and contain significant exposure levels of free silica.

Controls for Brazing Hazards

- Local ventilation control is necessary in operations where toxic metal fumes may be generated from the brazing components or from parts plated with cadmium or other toxic metals.
- Maintenance activities on the furnace and periodic replacement of the insulation material (usually refractive ceramic fibers) may result in exposure to free silica.

Welding

WELDING PROCESSES

- Welding is a process of joining metals in which coalescence is produced by heating the metals to a suitable temperature.
- Welding is classified as pressure, non-pressure, or brazing. Non-pressure welding fuses and melts metals together and requires special attention.

Non-pressure Welding

- Metal is vaporized and condenses to form a fume in the 0.01 to 0.1 μm particle size range. The respirable metal fume sources are the base metal, the metal coating on the work piece, the electrode, and the fluxing agents associated with a particular welding system.

- Depending on the welding process, a range of gases and vapors (ozone and nitrogen dioxide) may be generated.
- The electromagnetic radiation emitted depends on the welding procedure, inert gas, and base metal.

Shielded Metal Arc (SMA) Welding

- The non-pressure fusion welding process is commonly called "stick" or "electrode welding." This is an arc-welding process that produces coalescence of metals by heating them with an arc between a covered metal electrode and the work. Shielding is obtained from decomposition of the electrode covering. Pressure is not used and filler metal is obtained from the electrode.
- Metal fumes may contain iron oxides, manganese oxide, fluorides, silicon dioxide, titanium, nickel, chromium, molybdenum, vanadium, tungsten, copper, cobalt, lead, and zinc. Silicon dioxide is generated from the electrode coating.

Metal Fume Exposure

- Potential health hazards of SMA welding depend heavily on the metal being welded and the composition of the welding electrode.
- Stainless steel, the most common metal welded on, generates iron oxide. The hazard from exposure to iron oxide fume appears to be limited. The fume does cause a benign pneumoconiosis (known as siderosis) due to deposits of iron oxide particulates in the lungs.
- Concentrations of welding fumes to which an operator is exposed depend on:
  - Welding conditions (current density).
  - Arc time (may vary from 10 to 30% of the shift).
  - Power configuration (AC or dc).
  - Straight or reverse polarity.
- Exposure also depends on the local exhaust ventilation present, and whether the welding is done in an open, enclosed, or confined space. Air samples should be taken inside the welder's helmet, which range from 1/3 to 1/6 of the outside concentrations.

Gases and Vapors

- SMA welding may produce nitrogen oxides in significant concentrations. However, this is not normally a problem in open shop welding. The arc also fixes (makes stable) ozone, but the levels are not significant.

Radiation Hazards of SMA Welding

- The radiation generated by SMA welding covers the spectrum from the IR-C range of wavelengths to the UV-C range.
- The acute condition known to welders as "arc eye," "sand in the eye," "flash burn," or "welder's eye" is due to exposure in the UV-B range. The radiation in this range is completely absorbed in the corneal epithelium of the eye and causes a severe photokeratitis. It causes intense pain five to six hours after exposure to the arc. The condition clears within 24 hours. Welders can protect themselves against this by using a welding helmet with properly shaded lenses.
- Skin erythema or reddening may also be induced by exposure to UV-C and UV-B.

Low Hydrogen Welding

- Low hydrogen electrodes are used with conventional arc welding systems to maintain a hydrogen-free arc environment for critical welding tasks on certain steels.

Hazards of Low Hydrogen Welding

- Low hydrogen welding on low and high steel alloys produces oxides of nickel, chromium, molybdenum, and copper in addition to the normally observed oxides of iron, silicon, and manganese.
- Airborne fluorides are also present. Welding of this type without adequate local exhaust ventilation may result in nose and throat irritation and chronic nosebleeds.

Submerged Arc Welding

- In submerged arc welding, shielding of the arc from the atmosphere is accomplished by covering the weld with a granular, fusible flux. The granular flux is fed onto the metal ahead of the arc path, sintering (forming a homogeneous mass by heating without melting) to form a molten slag cover over the weld metal.
- Metal fume concentrations are lower with this type of welding process due to the blanketing action of the flux. The arc is maintained beneath the flux without sparks, smoke, or flash.

Resistance Welding

- In resistance welding, an electric current is passed through work pieces held together with pressure. There is localized heating at the contact point, and the metal coalesces.
- Various welding processes are based on this technique, including seam, spot, projection, and flash welding. No flux or filler is added. It is used widely on light sheet metal fabrications. Hazards are minimal.

Gas Tungsten Arc (GTA) Welding

- Gas tungsten arc welding (GTA), sometimes referred to as tungsten inert gas welding (TIG), is commonly used to weld aluminum, magnesium, nickel alloys, and other reactive metals.

- This welding process produces coalescence of metals by heating them with an arc between a tungsten (non-consumable) electrode and the work. Shielding is obtained from a gas or gas mixture. Pressure may or may not be used, and filler metal may or may not be used.

- The welding fume concentrations in GTA/TIG welding may be lower than manual stick welding if light gage material is being worked. The specific alloy and the filler wire in use must be identified to best assess the potential welding fume components and concentrations.

- The electromagnetic radiation from the GTA/TIG welding arc, which includes UV-B and C (especially in the 290 nm range), will produce skin erythema and photokeratitis. Energy below UV-C is most important in fixing oxygen as ozone.

Plasma Arc Welding

- In plasma arc welding, the welding head sends a flow of gas (such as argon) through an orifice under a high voltage gradient, resulting in an ionized gas stream. Mechanical and electromagnetic force produces arc temperatures greater than 60,000° F. This process is also used for cutting and metallizing. Hazards include noise, ozone, nitrogen dioxide, and metal fume exposure.

- Plasma welding may result in major skin and eye exposures requiring special clothing and eye protection. This type of welding usually requires a #14 shade and full chrome leather clothing, while manual SMA welding only requires a #10 shade. Intense UV-B radiation results in high ozone generation rates and fixing nitrogen dioxides.

- Noise may range from 110 to 120 dBA.

Gas Metal Arc (GMA) Welding

- Gas metal arc welding (GMA), sometimes referred to as manual inert gas welding (MIG), is commonly used to weld aluminum, copper, magnesium, nickel alloys, and titanium, as well as steel alloys.

- Gas metal arc welding is a process that coalesces metals by heating them with an arc between a consumable continuous filler metal electrode and the work. The shielding is obtained entirely from an externally supplied gas or gas mixture.

- The intensity of radiation emitted in GMA is similar to TIG welding and an order of magnitude greater than SMA welding.

- Additional hazards associated with UV radiation include the degradation of chlorinated hydrocarbons in the presence of welding fumes. Chlorinated hydrocarbon vapors are decomposed by UV emissions from the MIG arc, forming chlorine, hydrogen chloride, and phosgene. The solvent is degraded in the UV field and not directly in the arc. GMA/MIG welding may produce higher concentrations of phosgene than GTA/TIG welding under comparable operating conditions.

Gas Welding

- In gas welding, the heat of fusion is obtained from the combustion of oxygen and one of several gases, including acetylene, methylacetylene-propadiene (MAPP), propane, butane, and hydrogen. The flame heats the work piece and a filler rod is manually fed into the joint.

- Used widely for light sheet metal work and maintenance activities.

- Hazards are minimal compared to arc welding techniques. Metal fumes originate from the base metal, filler metal, and the flux. Since gas or torch welding is conducted at temperatures lower than those required for arc welding processes, excessive concentrations of metal fumes are seldom found, except when using lead, zinc, and cadmium.

- Gas welding emits a different range of electromagnetic radiation than arc welding. The principal emissions are in the visible and IR-A, IR-B, and IR-C wavelengths and require the use of lightly-tinted goggles.

Metal Fume Exposure

- Fume concentrations originate from the base metal, metal coatings, and welding electrodes.

- Beryllium, one of the most toxic alloying metals, is added to copper. As a result, special precautions are warranted when welding and brazing copper.

- Welding or cutting on work pieces that have metallic coatings may be especially hazardous. Lead-based paints, for example, generate lead fumes.

- Dipping in hot zinc dip galvanizes steel. Welding on galvanized metal can easily exceed the zinc TLV and result in the classic "Metal Fume Fever" symptoms. Symptoms resemble the flu, with headache, nausea, fatigue, and weakness. Symptoms appear shortly after exposure and dissipate when exposure ends.

- Local exhaust ventilation may be required in some situations, depending on the welding process, the base metal, the general ventilation systems in place, etc.

17 Environmental Issues

Environmental issues are becoming more and more important to industrial hygienists. Understanding the aspects of a variety of environmental regulations—for example, the handling, storage, and disposal of hazardous waste, air pollution, its formation, dispersion, and air sampling for air pollutants—plays a much larger role for the industrial hygiene professional today than in the recent past.

Toxic Substances and Hazardous Waste Management Operations

TOXIC SUBSTANCES CONTROL ACT (TSCA)

- Congress enacted the Toxic Substances Control Act (TSCA), PL 94-469, in 1976.
- TSCA provides the Environmental Protection Agency (EPA) with the authority to require testing of chemical substances entering the environment and to regulate them when necessary (referred to as pre-manufacture notification (PMN)).
- Regulatory actions include toxicity testing and environmental monitoring.

TSCA Title I—PCBs

- Title I regulates polychlorinated biphenyls (PCBs), and requires the EPA to take specific steps to control the risks they pose.

TSCA Title II—AHERA

- Title II, the Asbestos Hazard Emergency Response Act (AHERA), was enacted in 1986 and amended in 1988. Title II allows the EPA to set standards for responding to the presence of asbestos in schools.
- AHERA established responses based on the physical condition of asbestos. Schools were required to inspect for asbestos-containing material and develop a management plan for the removal of such material.
- AHERA requires asbestos contractors and analytical laboratories to be certified. It also requires schools to use certified people for asbestos work.
- AHERA was later amended to extend the training and accreditation requirement to include inspectors, contractors, and workers performing asbestos abatement work in all public and commercial buildings.

TSCA Title III—Radon

- The basic purpose of Title III, Indoor Radon Abatement, is to provide financial and technical assistance to states that support radon monitoring and control. The Act does not require monitoring or abatement.

- As required by Title III, the EPA has updated its pamphlet on radon, developed model construction standards for controlling radon levels in new buildings, and provided technical assistance to states.

TSCA Title IV—Residential Lead-Based Paint Hazard Reduction Act

- The goal of Title IV, the Residential Lead-Based Paint Hazard Reduction Act, is to reduce the risks to young children who are exposed to lead-based paint in their homes.

- Consistent with its goal, Title IV has encouraged the development of lead inspection and hazard abatement services in the private sector.

- The EPA was directed to develop definitions of lead-contaminated dust, soil, and paints.

- Title IV specifies requirements for accreditation of training programs for lead abatement work and criteria to evaluate the effectiveness of commercial products used to detect or reduce risks associated with lead-based paint. Title IV outlined certification requirements for laboratories performing such analyses.

- The EPA was required to conduct a study of lead hazards due to renovation and remodeling activities that may disturb lead-based paint, and promulgate guidelines for the renovation and remodeling of buildings or other structures when these activities might create a hazard.

RESOURCE CONSERVATION AND RECOVERY ACT (RCRA) OF 1976

- The Resource Conservation and Recovery Act (RCRA) established the federal program regulating solid and hazardous waste management. It amends an earlier act of 1965, the Solid Waste Disposal Act.

- RCRA expanded the federal government's role in solid waste disposal management, with an emphasis on hazardous waste disposal.

- RCRA prohibits the practice of open dumping.

- RCRA provides extensive federal aid through grants to state and regional agencies for solid waste planning and information programs.

RCRA—Cradle to Grave

- RCRA's cradle to grave program established requires that custody and responsibility move with a waste material from the generator and transporter to its final disposal site.

- The waste generator never loses liability for the waste created.

- The 1984 amendments to RCRA required small-quantity generators (those producing between 100 and 1000 kg per month) to meet certain procedural standards.

RCRA—General Waste Management Requirements

- The general waste management requirements of RCRA include proper notification and recording of hazardous waste activities, along with adequate packaging, labeling, and manifesting of wastes for shipment offsite.
- A RCRA permit is required for treatment, storage, and disposal (TSD) of hazardous waste onsite or offsite.
- Standards for TSD facilities include facility management plans, preparedness and prevention of emergencies and releases, contingency plans, operating records and reports, groundwater protection for land disposal facilities, and closure and post-closure plans with financial responsibility assurance.

RCRA—Changes Following the 1984 Amendments

- Later amendments made TSD facilities responsible for assessing human exposure to current and past waste management operations and for necessary corrective actions to remediate releases of hazardous substances to the environment.
- RCRA mandates the development of safe handling procedures, worker training, and emergency response measures. Training must be documented for review by the EPA.
- RCRA required that land disposal of specified highly hazardous wastes be phased out between 1986 and 1990.
- Technological standards were set for new landfills, mandating double liners, a leachate collection system, and groundwater monitoring.
- Under these amendments, states were encouraged to assume the EPA's hazardous waste program.

RCRA—Federal Facility Compliance Act

- RCRA's Federal Facility Compliance Act of 1992 allows the states, the EPA, and the Department of Justice to enforce the provisions of RCRA against federal facilities. Federal departments and agencies are now subjected to injunctions, administrative orders, or penalties for noncompliance.

COMPREHENSIVE ENVIRONMENTAL RESPONSE, COMPENSATION AND LIABILITY ACT (CERCLA) AND SUPERFUND AMENDMENT AND REAUTHORIZATION ACT (SARA)

- In 1980, Congress enacted the Comprehensive Environmental Response, Compensation, and Liability Act (CERCLA), PL 96-510.
- The Act established the Superfund program. Superfund manages emergencies at uncontrolled waste sites and clean up at these sites.
- In 1986, Congress reauthorized CERCLA as the Superfund Amendment and Reauthorization Act (SARA) of 1986, PL 99-499.
- The goals of the amendment were to provide additional funding and additional provisions, including underground storage tanks, emergency planning, risk assessment, community right-to-know, research, development, demonstrations, and training.
- These regulations further included requirements that owner/operators of leaking underground storage tanks undertake corrective action to protect human health and the environment.
- SARA also established a comprehensive federal program to promote various research, development, demonstration, and training activities.

AGENCY FOR TOXIC SUBSTANCES AND DISEASE REGISTRY (ATSDR)

- CERCLA created the Agency for Toxic Substances and Disease Registry (ATSDR) in the Public Health Service. ATSDR was created to carry out the health-related activities in the CERCLA act.
- EPA and ATSDR prepared a list of at least 275 of the hazardous substances most commonly found at the National Priority List (NPL) sites.
- The agency prepared toxicological profiles of these substances. Where information on a substance was lacking, ATSDR conducted its own research.
- ATSDR was required to perform a health assessment at each facility on the NPL list. ATSDR provides consultations to the EPA and state and local officials on health issues related to hazardous substances.

SUMMARY OF RCRA

- The RCRA amendments and cleanup requirements of CERCLA legislation forced the total accountability for current and past waste management practices into the limelight.
- Penal and financial liabilities for waste management practices are directly proportionate to how much a facility is affected by these requirements and remedial actions.

- Companies affected by the OSHA Hazard Communication Standard are also subject to the EPA Hazardous Chemical Reporting rules under Title III of Superfund.

- Affected companies' facilities must submit either copies of the MSDSs they prepare for OSHA compliance, or a list of all chemicals for which MSDSs are required, to the state emergency response commission, the local emergency planning committee, and the local fire department.

- Covered facilities must also submit emergency and hazardous chemical inventory forms to the same state and local authorities. Information on the daily amounts and chemical locations (designated Tier I information) must be submitted annually. The more detailed Tier II information is submitted on request.

HAZARDOUS WASTE OPERATIONS AND EMERGENCY RESPONSE (HAZWOPER)

- The treatment and disposal of hazardous wastes under RCRA and CERCLA creates a significant risk to the safety and health of employees who work in treatment and disposal operations. Skin contact, skin absorption, and inhalation pose the most significant risks to these employees.

- Employee exposure also occurs when employees respond to hazardous substance or hazardous waste emergencies.

- These conditions prompted OSHA to pass the Hazardous Waste Operations and Emergency Response (HAZWOPER) standard, 29 CFR 1910.120.

HAZWOPER—Applicability

- The HAZWOPER regulation addresses three different worker populations:
  - Employees engaged in the clean up of uncontrolled hazardous waste sites under CERCLA, including all Superfund sites.
  - Those employees engaged in operations involving hazardous waste treatment, storage, and disposal (TSD) under RCRA.
  - Those employees engaged in emergency response operations for releases or substantial threats of releases of hazardous substances and post-emergency response operations.

Site Safety and Health Program

- HAZWOPER requires a site safety and health program to be developed before work begins and to act as an extension of the safety program in place.

- The site safety and health plan facilitates coordination and communication of safety and health issues among personnel. It requires planning and implementing job training and orientation. It requires identifying and controlling work site hazards and the means for monitoring program effectiveness.

Emergency Response Plans

- HAZWOPER requires an emergency response plan be established and coordinated with local authorities.

Incident Command System (ICS)

- An incident command system (ICS) is required by this regulation, as an organized approach to effectively control and manage operations at an emergency incident.

- An ICS establishes a command post. The most senior officer usually assumes control and delegates tasks to handle specific situations.

Site Safety and Control Plans

- A comprehensive site safety and control plan must include:
  - Summary analysis of hazards on the site and a risk analysis of those hazards.
  - A site map.
  - Site work zones.
  - Use of the buddy system.
  - Site communications.
  - Command posts.
  - Standard operating procedures (SOPs).
  - Safe work practices.
  - Medical assistance and triage areas.

- Site safety and control plans should be part of the emergency response plan

Personnel Protection

- Personal protective equipment (PPE) should be used consistent with the type of situation encountered at a hazardous waste site or emergency response operation. HAZWOPER requires a written policy for the use of PPE.

Medical Surveillance

- Medical surveillance is essential to assess and monitor workers' health and fitness for employment in hazardous waste operations and during the course of work.

- Medical surveillance must include emergency and other treatment as needed. Adequate records must be maintained for future reference.

HAZWOPER TRAINING REQUIREMENTS

- HAZWOPER training varies with the type of work the individual will perform at the site. Training is most complex for individuals who are hazardous materials specialists, and becomes less detailed for individuals who will not respond to emergencies or work directly with hazardous materials.

Hazardous Waste Workers

- Hazardous waste workers should receive:
  - Information on the safety and health hazards expected to be encountered.
  - Effective control measures for those hazards.
  - Effective monitoring procedures for characterizing exposure levels.
  - Components of an effective employer's safety and health program.
  - Hands-on training with PPE and clothing expected to be used.
  - The OSHA standard.

Supervisory Training

- Supervisors receive training in their responsibilities under the safety and health program. They learn about the spill containment program, the PPE program, the medical surveillance program, the emergency response plan, and other areas.

Security and Support Personnel Training

- Security personnel must receive training and should thoroughly understand the incident command system (ICS).

- Support personnel (such as backhoe drivers) should be aware of the hazards and be provided with protective equipment.

Training of Technical Experts

- Technical experts who may respond to a situation should have training on an annual basis. It should include the care and use of PPE and respirators, knowledge of the ICS and their relationship to it, and information to keep them current in their respective fields.

Responder Training

- Responders to hazardous materials incidents should receive training on the care, use, and/or testing of chemical protective clothing, including totally encapsulating suits, the medical surveillance program, and the standard operating procedures (SOPs), as well as the additional information above.

Air Pollution

AIR POLLUTION REGULATIONS

- The Clean Air Act of 1955 was the basis of all subsequent legislation. It has been modified in a piecemeal fashion.

- The Act was amended in 1960 to include motor vehicle pollution. It was also amended in 1963 to include research on low sulfur fuels and to develop air quality criteria.

- The Air Quality Act replaced it in 1967, but was only applied to areas deemed to have poor air quality.

- When the EPA was created in 1970, the Act was expanded to cover the entire country.

- The New Source Performance Standards (NSPSs) were added in 1977.

- The Clean Air Act Amendments were added in 1990.

National Ambient Air Quality Standards (NAAQS)

- The National Ambient Air Quality Standards (NAAQS) established national primary standards for human health and secondary standards for the environment in 1970. CO, NO_2, SO_2, O_3, Pb, and PM-10 are regulated as "criteria pollutants."

- Geographic areas were designated as "attainment" or "non-attainment" areas based on their compliance with the standards.

Prevention of Significant Deterioration (PSD) Program

- The Prevention of Significant Deterioration program (PSD) is designed to keep attainment areas in attainment. It set emission standards for specific industries. It may require the use of "best available control technology" (BACT).

New Source Review (NSR) Program

- The New Source Review (NSR) is designed as a review process for any new sources in non-attainment areas. It could require "offsets" of emissions from other sources. It must comply with the "lowest achievable emission rate" (LAER).

National Emission Standards for Hazardous Air Pollutants

- The National Emission Standards for Hazardous Air Pollutants (NESHAP) set standards for emissions of toxic chemicals from specific industries.

- Between 1971 and 1984, eight standards were set for asbestos, beryllium, coke oven emissions, inorganic arsenic, mercury, radionuclides, vinyl chloride, and benzene.

New Source Performance Standards

- The New Source Performance Standards (NSPSs) set emission standards for both criteria and specific designated pollutants for new, modified, or reconstructed sources in specific industries.

1990 Clean Air Act Amendments

- The 1990 Clean Air Act Amendment was US legislation that amended the 1970 Clean Air Act. It consists of eight titles regulating everything from acid rain and motor vehicle emissions to ozone depleting chemicals.

- The Clean Air Act of 1990 increased the list of hazardous air pollutants (also called air toxics) from eight to 189. It requires major sources to obtain a federally enforceable permit for the release of air pollutants.

Titles of the Clean Air Act Amendments

- Title I—Ambient Air Quality Standard.

- Title II—Mobile Sources (cars, emission tests, fuel economy ratings, etc.).

- Title III—Hazardous Air Pollutants (HAPS).

- Title IV—Acid Rain.

- Title V—Federal Operating Permits (Very restrictive. Describes every operation that emits. HAPS restrict changes to permits).

- Title VI—Ozone (stratospheric ozone protection).

- Title VII—Enforcement.

- Title VIII—Miscellaneous Provisions.

WHAT IS AIR POLLUTION ?

- Air pollution is defined as the presence in the atmosphere of one or more air contaminants or combinations thereof in such quantities and of such duration that they are, or may tend to be, injurious to human, plant, or animal life, or property, or that may interfere with the comfortable enjoyment of life, or property, or the conduct of business or other human activities. These pollutants include gases, fumes, vapors, aerosols, and particles.

Factors to Consider in Air Pollution

- Critical factors to consider when evaluating air pollution problems include:
  - The atmospheric conditions that influence the dispersion or concentration of substances emitted into the atmosphere and the duration of exposure to these substances.
  - The effects of exposure to such concentrations over time.
  - The activities and processes that generate and emit substances to the atmosphere.
  - The extent to which these activities and processes may be modified, curtailed, or treated to reduce or eliminate emissions.

Air Pollution Causes

- The single major cause of air pollution is the internal combustion engine of automobiles. Gas is never completely burned in the engine of a car, just as coal is never completely burned in the furnace of a steel mill.
- These products of incomplete combustion (termed primary pollutants) are released into the atmosphere and may undergo a series of chemical reactions in the presence of sunlight, forming secondary pollutants, such as peroxyacteylnitrate (PAN, a component of smog).

Primary Pollutants

- Primary pollutants are those emitted directly from the source.
- Primary atmospheric pollutants include:
  - Inorganic primary pollutants (e.g., acid and alkali gases, such as sulfur dioxide (SO_2), hydrogen sulfide (H_2S), nitrogen oxides (NO_x,), ammonia (NH_3), hydrogen fluoride (HF), carbon dioxide (CO_2), carbon monoxide (CO), and ozone (O_3).
  - Organic primary pollutants (e.g., aliphatic and aromatic hydrocarbons, epoxides and peroxides of hydrocarbons, organic products of photochemical reactions, aldehydes, ketones, alcohols, acids, ethers, and esters).
  - Miscellaneous coarse dust (e.g., silica, alumina clay, fly ash, iron oxides, smoke and tarry material, polynuclear aromatic hydrocarbons (PNAs), carbon, and soot).

Secondary Pollutants

- Some secondary pollutants are formed as primary pollutants undergo change in the atmosphere. Others are naturally occurring.
- Secondary pollutants include:
  - Inorganic secondary pollutants (e.g., alkali metal compounds, light and heavy metals, nitrates, oxides, carbonates, chlorides, and fluorides, such as sulfuric acid (H_2SO_4), nitric acid (HNO_3), and ozone).
  - Organic secondary pollutants (e.g., oil and organic materials).
  - Viable particles (e.g., pollens, microorganisms, and insects).
  - Natural radioactive sources (e.g., igneous rocks and soils containing uranium, thorium, actinium series, and radon gas).

Acute Health Effects

- The most obvious acute health effects of air pollution have been shown in a few severe historic episodes. Most of these major episodes are related to atmospheric thermal inversions.
- Short-term exposures cause acute irritation of the respiratory tract, mild to severe cough, and affect the elderly who may have pre-existing the heart or respiratory system conditions.

Chronic Health Effects

- Chronic health effects include:
  - An increase in obstructive pulmonary disease and pulmonary emphysema. Both result in shortness of breath. SO_2, smoke and cigarette smoke are the most common causes.
  - Lung cancer caused by smoke, PNAs, and asbestos. The incidence of lung cancer is elevated in urban areas.
  - An elevated heavy metal content in blood.
  - An increase in carboxyhemoglobin.
  - Reduced lung functions.

Plant Health Effects

- Plant health is often adversely impacted by air pollution. Hazardous pollutants affecting plants include:
  - Sulfur dioxide (SO_2)—Produces necrosis (acute exposure) and chlorosis (chronic exposure).
  - Fluorides—Gaseous fluorides can be absorbed by leaf tips and result in chlorosis.
  - Oxidants—Peroxyacetylnitrate (PAN) in Los Angeles smog results in chlorosis and glazing on the underside of leaves.
  - Ozone—Causes stipple, dark marking, and flecking, as well as tight colored lesion necrotic patches.

HISTORIC AIR POLLUTION EVENTS

- Several significant air pollution incidents helped establish the effects of air pollution on human health. They also played a role in the establishment and refinement of US air pollution regulations.

Denora, PA, 1948

- In Denora, Pennsylvania, emissions from a large steel mill and a zinc reduction plant, combined with a thermal inversion, resulted in 20 deaths and approximately 6,000 illnesses in 1948. Denora is located in a valley. The inversion caused very high air pollution levels and concentrations.

- No specific findings on the autopsies alerted people to public health hazards of chemical air pollutants.

Los Angeles, CA, 1950

- A widespread awareness of pollution began in Los Angeles in the early 1950s.

- The Los Angeles basin is ringed for the most part by high mountains. As cool air sinks from these mountains, it is heated until it accumulates as a warm layer that rises above the cooler air from the Pacific Ocean.

- This creates a temperature inversion, with the heavier cool air confined to the surface. Pollutants also become trapped at surface levels.

- Because of air-circulation patterns in the Los Angeles Basin, polluted air merely moves from one part of the basin to another.

London, 1952

- In 1952, London attracted worldwide attention when 4,000 people died, leaving no doubt that air pollution can kill. Conditions were fog, a temperature inversion, and smoke and SO_2 from domestic heating fires.

Bhopal, India, 1984

- Union Carbide, the third largest American chemical company, had a facility in Bhopal that accidentally released the poisonous gas methyl isocyanate into the community. More than 2500 men, women, and children were killed when the gas swept through the city.
- The incident triggered a new emphasis on emergency contingency planning and emergency response worldwide. The incident seriously damaged the company's reputation and brought it massive litigation problems.
- A similar incident occurred shortly thereafter at Union Carbide’s plant in Institute, West Virginia. However, operators there were able to shut down the system before any injuries occurred.

AIR POLLUTION CHEMISTRY

- It is important to understand the chemistry of air pollution in order to be able to measure emissions and implement engineering controls.

Smog Formation

- Smog is formed in the atmosphere through a chemical reaction. The principal chemical component of smog is peroxyacetylnitrate (PAN), which is formed in the atmosphere in the presence of organic contaminants.

$$\frac{(NO_x)\ (\text{Organic Concentration})\ (\text{sunlight intensity})\ (\text{Temperature})}{(\text{windspeed})\ (\text{inversion height})}$$

- Smog may appear brown when it contains high concentrations of nitrogen oxides (NO_x), or blue-gray when it contains large amounts of ozone.

Acid Rain

- When sulfur dioxide emissions from electric power plants combine with particles of water in the atmosphere, they form sulfuric acid, carbonic acid and nitric acid. This falls to the ground as acid rain or acid snow.

- The acidity or basicity of liquids, including rainfall and snow, is measured by pH. Normal rainwater has a pH of 5.5. The National Center for Atmospheric Research has recorded storms in the northeastern United States with a pH of 2.1, an acidity equivalent to that of vinegar or lemon juice.

- Acid rain is blamed for the deaths of thousands of lakes and streams. These water bodies have absorbed so much acid rain that they can no longer support algae, plankton, and other aquatic life.

- The effects of acid rain are not limited to geographic boundaries. Prevailing winds also carry the acid around the globe.

- Nor are the destructive effects of acid rain limited to the natural environment. Structures made of stone, metal, and cement have also been damaged or destroyed. Great monuments, including the cathedrals of Europe and the Coliseum in Rome, have shown signs of acid rain deterioration.

CFCs and Ozone Depletion

- Another form of air pollution comes from a variety of chlorofluorocarbons (CFCs). These chemicals are used for many industrial purposes, ranging from cleaning solvents to refrigerant gases found in air conditioners.

- CFCs released into the atmosphere are decomposed by ultraviolet light. This chemical action releases chlorine atoms that react with ozone to form new molecules. Theoretically, this process depletes atmospheric ozone. At ground level, ozone is a threat to our lungs. But in the upper atmosphere, it works as a protective shield against ultraviolet radiation from the sun.

- Less ozone in the stratosphere may increase exposure to ultraviolet radiation at ground level, causing damage to crops and the eye, and increasing skin cancer.

- Use of CFCs has been phased out by legislation. However, debate continues regarding the facts about ozone depletion and its relationship to CFCs. Finding substitutes for CFCs has been a difficult task for chemical users and producers.

Meteorology

- How pollution travels in the atmosphere depends on meteorology. Pollutant travel or behavior in the air is often described by air pollution plumes. These plumes include looping, coning, fanning, lofting, and fumigating air currents.

- Temperature lapse rates also impact how pollution travels in the atmosphere:
  - Adiabatic—Unsaturated air as it increases in height in the atmosphere decreases in temperature @ ~ 10°C/km.
  - Super adiabatic—If air is warmer and less dense, the decrease in temperature occurs faster with height than the adiabatic rate.
  - Inversion—Negative lapse rate (temperature increases with height, an extreme case of atmospheric stability).
  - Isothermal—No temperature change occurs with height.

Air Flow Obstacles

- Meteorology impacts the movement of air pollution in the atmosphere. Outlined below is how eddy currents impact plumes. This information is useful for industrial hygienists when implementing air pollution control measures.

ΔH = Effective stack height (stack plus plume height)
Δh = Plume height
h = Stack height

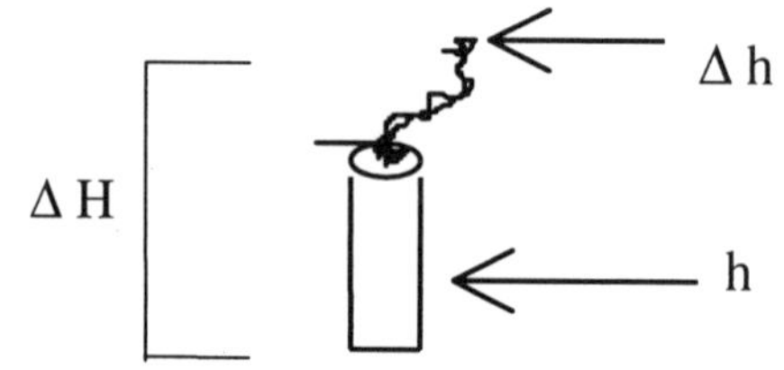

↑ Volume or mass emission rate (faster, higher) ↑ Δ H
↑ Effluent temperature becomes less dense so ↑ Δ H
↓ Ambient air temperature ↑ Δ H
↑ Ambient air temperature lapse rate ↑ Δ H
↓ Effluent density (lighter) ↑ Δ H
↑ Effluent velocity (faster) ↑ Δ H
↑ Wind speed ↓ Δ H

AIR SAMPLING

- Industrial hygienists are frequently called upon to perform air pollution monitoring. The instruments and methods used are, in many cases, identical to those used for workplace air sampling.

Standards

- Primary standards: Involve direct measurement of volume on the basis of the physical dimensions of an enclosed space. Examples: Bubble meter, spirometer, and gasometer.

- ❑ Secondary standards: Reference instruments or meters that trace their calibration to primary standards and are shown to be capable of maintaining their accuracy with reasonable handling and care in operation.

- ❑ Examples: Rate meter and wet-test meter.

Particle Collection Devices for Air Pollution

- ❑ Particle collection devices for air pollution include many devices used for industrial hygiene air sampling. These include:
  - ❑ Hi-Volume (hi-vol), which collects 5-15 cfm (weigh filter afterward).
  - ❑ Two-stage dust sampler.
  - ❑ Cyclone.
  - ❑ Filter tape samples—used in community air pollution measurement. Checks CO, air traffic, and opacity readings.
  - ❑ Filters.
  - ❑ Particle counters.
  - ❑ Impingers—use liquids.
  - ❑ Impactors—cascade, dry.
  - ❑ Anderson sampler—works like a cascade impactor. Collection media can be a culture media.
  - ❑ Electrostatic precipitator.
  - ❑ Thermal precipitator.

Gas and Vapor Collectors

- ❑ Collectors for gaseous and vapor air pollutants include:
  - ❑ Air displacement devices.
  - ❑ Impingers.
  - ❑ Activated charcoal.
  - ❑ Silica gel.

Source Testing

- ❑ Source testing or stack sampling of particulates uses a special pitot tube and considers effluent gas velocity.

- ❑ Stacks must be sampled isokinetically. That is, sampling must be done at the same rate as the gas stream. If sample volume is too high, too many particles will be pulled through. If sample volume is too low, particles will bypass the sampler and subsequently cause an undersample.

- ❑ ORSAT is one piece of equipment used for stack sampling that measures the density of flue gas and compares it to standard air.

ENGINEERING CONTROLS FOR AIR POLLUTION

- ❑ Engineering controls may be process and fuel changes (i.e., substitution).

- Engineering controls also include local exhaust ventilation with particle collection or air cleaning devices. These include:
  - Filtration (such as baghouse filters).
  - Electrostatic precipitators.
  - Cyclones.
  - Mechanical collectors (such as gravity settling chambers).
  - Wet collectors (scrubbers).
  - Absorption of gases.
  - Adsorption of gases.
  - Incineration.
  - Catalytic combustion.

Characteristics of Pollutants Used to Select Control Devices

- The characteristics of pollutants help to determine the appropriate control method or device. These characteristics include:
  - Loading.
  - Solubility.
  - Temperature of effluent.
  - Viscosity.
  - Density.
  - Sorbability.
  - Reactivity.
  - Combustibility.
  - Toxicity.
  - Particle size and shape.
  - Hygroscopicity.
  - Agglomeration.
  - Flow properties.
  - Electrical properties.
  - Humidity.

18 Hazard Communication

Hazard Communication is one of OSHA's most significant standards. The intent of the standard is to provide employees with information about the potential health hazards associated with exposure to workplace chemicals. Industrial hygienists must be very familiar with the Hazard Communication standard, its purpose, and requirements. The role of an industrial hygienist involves risk assessment, employee training, and material safety data sheet (MSDS) interpretation.

Hazard Communication (HAZCOM): The Employee's Right-to-Know

HAZARD COMMUNICATION (HAZCOM)

- The federal Hazard Communication (HAZCOM) standard, sometimes referred to as the Right-to-Know Law, 29 CFR 1910.1200, became effective May 1986.
- Hazard Communication is a performance standard. Performance standards allow employers to use different means to comply with the law.
- The law was enacted to prevent employees from unknowingly working with hazardous chemicals that may cause them immediate or delayed injury or illness.
- The intent of the standard is to provide employees with information about potential health hazards from their exposure to workplace chemicals.
- The standard mandates employee training, which educates employees on the risks of their work and the necessary precautions to minimize those risks.

Applicability

- The HAZCOM standard applies to chemical manufacturers, importers, and distributors of chemicals.
- The standard also applies to all employers with employees who are exposed to hazardous chemicals as part of their employment. Laboratories are not exempt.
- Only chemicals and materials considered hazardous are covered by this standard.
- The standard applies to any chemical known to be present in the workplace that employees may be exposed to under normal conditions of use or may be exposed to in a foreseeable emergency.

Written Hazard Communication Program

- It is an employer's responsibility to establish and implement a hazard communication program.

- This program must be a written program. The written program must describe how the requirements of the Hazard Communication Standard will be met.

- The written program must include a list of hazardous chemicals known to be present at the employer's site. This list serves as an index to the material safety data sheets (MSDSs). Material safety data sheets must be readily available to employees.

- The written program must include the methods the employer will use to inform employees of the hazards of routine and non-routine tasks and the hazards of chemicals in unlabeled pipes. Pipes do not have to be labeled under the standard, but it requires them to have a means of conveying to employees the hazards of any operations with pipes containing hazardous materials.

- OSHA inspectors will ask to see this written program and verification of its effectiveness during the opening conference of an inspection performed at the employer's site.

- The sections of the hazard communication standard most frequently cited by OSHA inspectors are those dealing with the lack of a written hazard communication program, lack of a training program, and lack of labels on hazardous chemical containers.

Chemical Evaluation and Inventory

- Only hazardous materials or chemicals are covered by this regulation. Hazardous materials include various paints, solvents, coolants, cutting oils, fuels, hydraulic lubricants, insulation products, epoxies, adhesives, dyes, and asbestos-containing products.

- The Hazard Communication standard requires that all chemicals be evaluated to determine their hazards.

- The chemical's manufacturer may determine whether a material is considered to be hazardous.

- If the National Toxicology Program (NTP) or International Agency for Research on Cancer (IARC) or other study declares the chemical a possible carcinogen, the chemical is considered hazardous and is affected by this regulation. The information must be specified in the MSDS, whether or not the reviewer or manufacturer agrees that it is a carcinogen.

- Once the chemical has been evaluated, and hazard information has been compiled, the standard requires that this information be transmitted to employers and employees through labels, MSDSs, and training.

- The standard requires that employers maintain an accurate inventory of the hazardous chemicals used on their premises.

- The inventory also serves as an index of the MSDSs that must be available to employees.

Labels

- The Hazard Communication standard requires employers to label all containers of hazardous materials used in the workplace.

- The purpose of the labeling requirements are:
  - To serve as an immediate visual warning that the chemicals in the container are hazardous.
  - To remind employees of what they have been taught about handling these chemicals.
  - To advise that more detailed information is available on the MSDS.

- When a chemical manufacturer ships a chemical to a user, the original container must be labeled.

- The label on the original container must include the following:
  - Identity of the chemicals contained.
  - Appropriate hazard warnings.
  - Name and address of the chemical manufacturer, importer, or other responsible party.

- In an industrial environment, chemicals are not always used in their original containers, especially when they are shipped in large fifty-five gallon drums. These hazardous materials are transferred to in-house containers.

- Once chemicals are transferred from their original containers, in-house containers must include the identity of the substance and appropriate hazard warnings.

- Employees are discouraged or prohibited from removing labels as long as the substance named on the label remains in the container.

- ❑ The label is the first indicator of a possible hazard. Employees are encouraged to refer to material safety data sheets for information not found on the label.

Material Safety Data Sheets (MSDSs)

- ❑ Material safety data sheets are written documents that provide extensive information on the chemical's components, as well as other useful information, such as the hazards of the chemical and protective measures to be taken when using them.
- ❑ These documents must contain specific information on when a chemical or chemical mixture contains a carcinogen.
- ❑ If a chemical is declared a possible carcinogen by the National Toxicology Program (NTP) or International Agency for Research on Cancer (IARC) or other study, the chemical is considered hazardous and that information must be specified on the MSDS, whether or not the reviewer or manufacturer agrees that the chemical is a carcinogen.
- ❑ Mixtures are assumed to have the same hazards as any of its chemical components comprising one percent or more of the total mixture. These chemicals and their hazards must be identified on the MSDS.
- ❑ A chemical mixture that poses a carcinogen hazard or contains a carcinogen in concentrations of 0.1 percent or more must be identified as such on the MSDS.
- ❑ MSDSs must be made readily available to employees and their designated representatives. Thus employees will always have access to detailed information on the identities and hazards of the substances they are exposed to in their work areas.
- ❑ MSDSs are required for each hazardous chemical in the workplace.

Training

- ❑ Training is required for all employees exposed to hazardous chemicals before their initial assignment to such work and whenever the hazard changes.
- ❑ Explanations of the requirements of the standard are a necessary component of training.
- ❑ Workplace operations where hazardous chemicals are present must be identified and discussed.

- Knowledge of the methods and observations used to detect the presence of hazardous workplace chemicals must be conveyed.

- How employees can determine hazards associated with chemicals in unlabeled pipes is a necessary component of training.

- Description of hazards associated with non-routine tasks is a significant component of training. This training is often omitted. This requirement applies to maintenance personnel who may be required to work in a hazardous environment for repairs and renovations. It also applies to outside contractors who enter the workplace for such repairs and renovations.

- Details about the measures employees can take to protect themselves against these hazards, including specific measures, must be included in training.

- The training must include an explanation of the labeling system in place.

- The training must include instructions on the location and use of material safety data sheets or material safety sheet databases accessible to employees.

- Training usually includes information about the health hazards of a material, covering such topics as carcinogenity, toxicity, reproductive effects, irritation, corrosiveness, sensitization, and systemic poisoning.

- Information about the physical hazards of a material or chemical are usually included in training. Physical hazards may be present in materials that are combustible, flammable, compressed gases, explosive, organic peroxides, oxidizers, pyrophoric, reactive, or water-reactive.

- Training may include a section-by-section review of the contents of a material safety data sheet.

- Training may be offered through computerized interactive training, classroom training, or other training methods.

Trade Secrets

- When a specific chemical identity is a bona fide trade secret, the chemical manufacturer, importer, or employer can withhold the specific chemical identity. The identity can be withheld as long as all other information about the chemical is disclosed, and it is noted on the MSDS that the specific chemical identity is being withheld because it is a trade secret.

- The standard requires disclosure of the trade secrets to health professionals, employees, or designated representatives who provide medical or other occupational health services to exposed employees.

- To obtain trade secret information, a written request is required:
  - Substantiating the need for the information.
  - Stating why other information would not satisfy that need.
  - Describing the means that will be used to keep the information confidential.

- The holder of the trade secret must disclose the information if these requirements are met.

- If the holder denies the request, the denial must be made in writing and provided within 30 days of the request.

- The denial must include evidence that the specific chemical identity is a trade secret, state why the request is being denied, and explain how alternative information may satisfy the health information request.

- The matter may then be referred to OSHA.

Recordkeeping

- Written plans of compliance and material safety data sheets must be kept current.

- Outdated sheets need not be kept unless they are subject to the requirements of 29 CFR 1910.20, Access to Employee Exposure and Medical Records.

- Under that rule, if an employer has no other record of exposure, the data sheets may have to be retained for 30 years.

19 OSHA Reporting, Inspections, and Citations

Most industrial hygienists will, at some point in their career, participate in the OSHA recordkeeping, inspection, and citation processes. Recordkeeping is no longer the sole responsibility of the safety engineer. Industrial hygienists must be familiar with specific injuries and illnesses considered recordable and understand what constitutes a recordable injury or illness. A company industrial hygienist may sample side-by-side with an OSHA industrial hygienist during an inspection to demonstrate repeatability and validity of the company's air sampling program.

OSHA Injury and Illness Reporting

OSHA REPORTING REQUIREMENTS

- Many US employers are required by OSHA to maintain accurate records of all work-related injuries and illnesses that occur on their premises.
- When an employee in the course of his or her work cuts a finger and needs sutures, the injuries is clearly recordable and must be logged on the OSHA recordkeeping and reporting forms. However, there are injuries and illnesses that are not as clearly defined.
- For example, an employee who nearly chokes to death on a chicken bone while eating in the employee cafeteria, or the employee who sprains an ankle by tripping on a rock in the company parking lot. Did these injuries occur in the course of employment, or did they just happen at the work site? Certainly not 'black and white,' these are the types of injuries that cause over- and under-reporting of workplace injuries.

Proposed Changes

- In early 1996, OSHA held hearings on its proposal for recordkeeping changes.
- The latest information can be obtained on the World Wide Web at http://www.osha.gov/ and accessing the "What's New" section.

Purpose of Changes

- The purpose of the proposed recordkeeping changes is to simplify processes and eliminate confusion.
- The proposed changes lift recordkeeping requirements from thousands of employers who have less than 10 employees. The proposal could also exclude many employers with 19 or fewer employees.
- Other proposed changes include smaller standard OSHA forms, an option for employers to keep records on computer, and the end of the distinction between an injury and an illness.

Exemptions

- Under the proposal, employers in the construction industry with 10 or fewer employees are exempt.
- Non-construction employers with 19 or fewer employees are exempt, unless the report shows a fatality or multiple hospitalization incidents.
- Auto and home supply stores would now have to comply with all recordkeeping requirements.

Forms

- Under the proposal, each employer will maintain an OSHA Injury and Illness Log and Summary—OSHA Form 300—which replaces the OSHA 200 log.
- Each employer must complete an OSHA Injury and Illness Incident Record, OSHA Form 301, formerly OSHA Form 101.
- Incidents must be recorded within seven calendar days of receiving information about the case.

What Qualifies as Recordable?

- A recordable injury must be work-related, a new incident, and must meet at least one of these criteria:
  - Results in death or loss of consciousness.
  - Results in days away from work, restricted activity, or job transfer.
  - Requires medical treatment beyond first aid.
  - Is a recordable condition listed in the proposal's Mandatory Appendix B.

Illness Reporting

- Currently, OSHA requires all occupational illnesses to be recorded.
- To encourage employers to more accurately track illnesses (thus improving the quality of the data), OSHA has proposed using mandatory Appendix B, a list of specific conditions with corresponding recording criteria.
- Examples of illness reporting include:
  - Elevated blood lead levels.
  - Hearing loss (a 15 dB or greater shift at 2000, 3000, and 4000 Hz).
  - Skin disorders lasting beyond 48 hours.
  - Asthma.
  - Asbestos-related disorders.
  - Bloodborne pathogens diseases.
  - TB and other infectious diseases.

First Aid

- Recordkeeping compliance comes down to being able to understand definitions.
- OSHA says an injury or illness is recordable if it requires medical treatment beyond first aid.
- First aid includes:
  - Visit(s) to a health care provider for limited observation.
  - Diagnostic procedures, including the use of prescription medications solely for diagnostic purposes, such as eye drops to dilate the eyes.
  - Use of non-prescription medications, including antiseptics.
  - Simple administration of oxygen.
  - Administration of tetanus or diphtheria shot or booster.
  - Cleaning, flushing, or soaking wound on skin surface.
  - Use of wound coverings, such as bandages, gauze pads, etc.
  - Use of any hot/cold therapy, except for musculoskeletal disorders.
  - Use of any totally non-rigid, non-immobilizing means of support, such as elastic bandages.
  - Drilling of a nail to relieve pressure for subungual hematoma.
  - Use of eye patches.
  - Removal of foreign bodies not embedded in the eye if only irrigation or removal with a cotton swab is required.
  - Removal of splinters or foreign material from areas other than the eyes by irrigation, tweezers, cotton swabs, or other simple means.

Work-Related Injuries

- Injuries and illnesses are considered to be work related if an event or exposure in the work environment either caused or contributed to the condition, or aggravated a pre-existing condition.
- Work-related also applies to injuries and illnesses sustained while employees are engaged in a required apprenticeship or vocational training, on break, in rest rooms or storage areas located on the employer's premises, performing work at home, and traveling on company business.

Non-work Related Injuries

- In the proposed regulation, OSHA much more clearly defines injuries and illnesses that are not considered work related.
- An injury or illness is not considered work related under the following conditions:
  - When individuals are on company premises as a member of the general public, not as a worker.
  - Symptoms are noticed at work but result solely from a non-work-related event or exposure outside the work environment.

- ❑ When workers voluntarily participate in wellness programs, medical activities, or recreational activities.
- ❑ When workers are eating, drinking, or preparing their own food in the workplace.
- ❑ When workers are doing personal tasks outside of normal working hours.
- ❑ When a worker, his or her family, or ex-spouse commits acts of violence unrelated to the worker's job. This includes intentionally self-inflicted injuries.
- ❑ When employees are entering or leaving company parking lots and access roads.
- ❑ If the worker was never engaged in any duty at work that could have placed stress on the affected body part, or was never exposed to any chemical or physical agent at work that could be associated with the observed injury or illness.
- ❑ When workers are involved in voluntary community or civic projects away from the company premises.
- ❑ If the case solely results from normal body movements, such as walking unencumbered, talking, tying a shoe, sneezing, coughing, etc., provided the activity doesn't involve a job-related motion and the work environment is not a contributing factor.
- ❑ If the worker experiences mental illness associated with post-traumatic stress.

New Cases

❑ A case is considered new if an injury or illness does not result from the recurrence of a pre-existing condition and if no new workplace incident or exposure occurs.

❑ A recurrence of a previous case is presumed to be new when it results from a new work situation, or if 45 days have elapsed since medical treatment, restricted work, or days away from work were discontinued and the last signs or symptoms were experienced.

Subcontractor Records

❑ Any controlling employer in the construction industry overseeing a project with an initial contract worth at least $1 million must maintain a separate injury and illness log of recordable cases sustained by subcontractor employees.

❑ Records do not have to be kept for subcontractors with less than 11 employees.

❑ Information for each recordable case involving a subcontractor employee includes the person's name, company, date of event, and a brief description of the injury or illness.

- This does not relieve subcontractors of their responsibilities to complete a year-end injury and illness summary, OSHA 301, and more indepth records.

The OSHA Inspection and Citation Process

COMPLIANCE INSPECTIONS

- Five mechanisms may initiate an on-site visit by an OSHA compliance officer:
  - Complaint.
  - General scheduling.
  - NIOSH Health Hazard Evaluation.
  - Fatality or catastrophic event.
  - Referral.

The Complaint

- Complaints received by OSHA are prioritized as follows:
  - By employees.
  - By the immediate families of employees.
  - By labor representatives.
  - Anonymous complaints.
- The complaint is rated as serious or non-serious, and formal or non-formal.
- Guidelines—such as the type of hazard, employees exposed, duration, etc.—are used to rate a complaint's seriousness.
- A formal complaint is one that is signed by the complainant.
- Formal serious complaints are always investigated.
- Some regional OSHA offices send letters to employers advising them of an informal complaint and request a reply to the allegation(s).
- Some OSHA inspectors will even telephone the employer. Receiving no response or a poor response is then grounds for an investigation.

General Schedule

- A system of "worst first" has been devised so that general inspections are performed at the workplaces that contain the greatest hazards. The list is created using standard industrial classification (SIC) numbers.
- If NIOSH has performed a health hazard evaluation (HHE) in a workplace and serious violations exist, OSHA is sent the final copy of the HHE for follow-up.

Fatalities

- All industrial fatalities are investigated after the receipt of a fatality report. The majority of these investigations are "safety related."
- 29 CFR 1904.8 provides that when reporting fatalities and hospitalization of employees:
    - Reports of a fatality or multiple hospitalization must be filed within eight hours (formerly 48 hours).
    - Incidents that result in three or more hospitalized employees (formerly five or more) are to be reported.

Referral

- Safety compliance officers may observe health situations they are not equipped to investigate. An inter-office referral is often made for an industrial hygienist to follow-up. However, some OSHA compliance officers are industrial hygienists. In some cases, an industrial hygienist may accompany the safety compliance officer during the inspection. A health inspection (including air sampling) may be conducted concurrently with the safety inspection.

Opening Conference

- The reason for the OSHA inspection is discussed in the opening conference. Management officials of the organization are usually present. If a labor union representative is present in the workplace, that representative is contacted.
- The scope of the visit is covered.
- The company's OSHA accident/injury records are reviewed, as well as their written Hazard Communication Program.

Walk-Through Inspection

- The walk-through inspection is usually limited to the complaint area, unless the company appears on the high hazard SIC list.
- A general schedule inspection covers the entire facility.
- Areas to be sampled are selected at this time.

First Closing Conference

- If it is necessary to conduct full shift sampling (full shift minus one hour), the OSHA compliance officer or industrial hygienist (IH) informs management during the first closing conference.
- Arrangements are made for the OSHA IH to return at the beginning of the shift.
- It is not necessary to disclose where (which process, area, etc.) the sampling is to be performed at this time.

- If it is determined by the OSHA IH that a violation does not exist, the inspection ends here.

Sampling

- Full shift sampling is conducted in accordance with the *Field Inspection Reference Manual* (FIRM).
- All sampling and analysis errors (SAEs) must be taken into account to fully document a violation.
- Sampling and analytical error factors for various compounds (as well as noise exposure) are published in the FIRM.
- SAE's provide a 95% confidence limit. Upper and lower confidence limits are calculated on a laboratory result before a violation is established.

The Closing Conference

- Results and conclusions are presented to the employer during the closing conference.
- The company industrial hygienist, environmental safety and health leader, and management officials of the organization are usually present.
- If a labor union representative is present in the workplace, that representative may be present at this meeting.

CITATIONS

- If a violation is established, it is classified as "serious" or "other" depending on the type of hazard and level of hazard measured.

General Duty Clause

- Section 5(a)(1) of the OSH Act, known as the General Duty Clause (GDC), is used when exposure to a hazard does not correspond to an OSHA standard, but is believed to constitute a serious health hazard.
- The General Duty Clause it is designed to protect all workers where there is no specific regulation in place.

Contests and Abatement

- After the citations are issued, the employer may contest the citations in writing to the OSHA Area Office within 15 working days.
- If penalties are assessed and there is no contest, payment must be made within the same 15 working days.
- Under the OSH Act, the employer has the right to meet with the Regional Administrator. This has been delegated to Area Directors (ADs).

- During the 15 days after citation(s) are issued, the employer can meet with the AD in an "informal conference" to discuss the citation(s). The AD has the authority to reach a settlement between OSHA and the company over issued citation(s). This mechanism has lowered the number of formal contests.

Abatement

- Abatement dates are assigned after discussion during the closing conference.
- The AD can change these dates during the informal conference.
- When the abatement date is almost due, the employer can request an extension, in writing, and justify his or her request.

Violation Definitions

- Serious: Violation can cause death or physical harm. Most IH issues are considered serious.
- Willful: Employer demonstrates an intentional violation of the OSH Act or an indifference to its requirements.
- Repeat: A violation cited previously for the same condition or activity. This may be the same condition or activity within same company or at another site or department. Repeated violations differ from willful violations in that they may result from an inadvertent, accidental, or ordinarily negligent act.
- Failure to Abate: A failure to abate situation exists when a situation or condition previously cited has never been brought into compliance and is noted on a later inspection.
- Egregious: Must be classified as willful and meet other criteria, such as worker fatality, catastrophe, high injury rate, extensive history of poor compliance, intentional disregard for safety, bad faith, and history of a large number of violations.
- Other-than-serious: A violation that will probably not result in irrevocable harm, such as a tripping hazard.
- De minimis: Minor violations, usually in administrative areas such as not posting the OSHA log.

20 Ethics and Management

Professional ethics has changed dramatically in the field of industrial hygiene. As industrial hygienists move from traditional technical and specialist roles into more generalist roles, management skills become more important.

Ethics and the Industrial Hygienist

PRIORITY SETTING

- ❑ Costs versus benefits must be considered in the industrial hygiene profession. Some believe that placing a price on human life is immoral — that any effort, even to save one life, is worthwhile.
- ❑ Some form of priority setting must be considered.
- ❑ Industrial hygienists rank priorities daily.
- ❑ Decisions may be small compared to national standards, yet many recommendations can result in reduced profits for employers and clients.
- ❑ Ranking is essential to differentiate between essential and desired solutions, just as OSHA ranks its citations in descending order from willful to de minimis.

INDUSTRIAL HYGIENE AS A PROFESSION

- ❑ Industrial hygienists must be able to rise above the fog of professional jargon and make their knowledge and expertise available to others.
- ❑ Industrial hygienists must enjoy the confidence and trust of the clients and employers who count on them for protection.

Prevention

- ❑ Physicians treat the sick, lawyers protect legal rights, and industrial hygienists prevent work-related and environmental accidents and illnesses.
- ❑ The professional industrial hygienist is attuned to prevention. Prevention saves lives and suffering. When properly implemented, it is also cost-effective and earns the respect and support of workers.

Sound Recommendations

- ❑ No industrial hygienist should expect that sound recommendations will always be accepted on merit alone, without proper explanation.
- ❑ Never consider "selling" your recommendations to clients and employers to be beneath your professional dignity.
- ❑ Prevention programs are often not adequately explained to management and as a result, end up being rejected.

- Rather than blame management, the bottom line, or the cash register, the industrial hygienist should consider:
  - If the industrial hygienist was the client or employer, would he or she pay for these programs?
  - Has full documentation been provided?
  - Has the situation been proposed as a crisis rather than as a factual, data-driven situation? Crisis management will not earn the credibility and respect necessary to influence prevention recommendations.
  - Has the art of influence been applied? Making good recommendations is a professional duty, but selling an idea is an art of influence.

Professional Integrity

- The best quality industrial hygienists have to offer is their professional integrity.

- Industrial hygienists must present the same findings and recommendations to both management and labor.

Management and Industrial Hygiene

TRADITIONAL MANAGEMENT

- Keeping abreast of management issues varies with each company an industrial hygienist works for. Traditional management is required for organizational control, planning, budgeting, direction and supervision. A pyramid structure is the normal environment, and individuals usually work for a single supervisor or manager.

MANAGEMENT TODAY

- Today's management has shifted to require:
  - Flat, horizontal organizations.
  - Matrix organizations, where individuals work for more than one person at a time.
  - More self-directed work groups.
  - Specialized work groups to plan budgets, overtime, goals, etc.
  - Team environments rather than individual responsibilities.
  - Total Quality Management (TQM), Six Sigma, and other program influences.
  - A re-engineering versus re-structuring emphasis.
  - A focus on generalization rather than specialization.

Line Versus Staff Organizations

- Line management is typically responsible for:
  - Directing the conduct of the business of the organization.
  - Getting the product built, shipped, and delivered to the customer.
  - Following agreed-to objectives.
  - Reporting variances from the business plan.
  - Developing and implementing corrective actions to stay on schedule.

- In forward-thinking companies, line management is also held accountable for the safety and health of its employees.

- Staff organizations are supporting functions enabling line management to run the enterprise. Industrial hygiene functions usually fall into the staff organization responsibility.

- Staff responsibilities include advice, counsel, and authoritative opinion on specialized areas, interpreting policies and practices, and establishing an implementation plan.

- Many staff functions are also now required of the industrial hygienist to follow agreed-to objectives, report variances to the plan, and develop and implement corrective actions to stay on schedule, within budget, etc.

LEADERSHIP CHARACTERISTICS

- Organizations are looking for managers who can lead: dynamic, flexible individuals capable of inspiring other people.

- Today's managers must be able to do more than the traditional manager who sets goals, directs, and controls. These managers must lead and be able to:
  - Take charge: Effectiveness depends heavily on credibility among the employees. That credibility is undermined if the leader fails to "walk the talk."
  - Set a clear agenda: New problems compete for attention and people pursue conflicting agendas. Because of the resulting blur, people head in many different directions. Stay abreast of new priorities and map out the game plan.
  - Show a sense of urgency: Show fire, energy, speed, and action, because employees will respond in kind.
  - Accept ambiguity and uncertainty: Organizations change very quickly. Their leaders change, and their goals change. A good leader must be able work under ambiguous conditions and not allow employees to wallow around and wait to hear the "new position" before accomplishing their goals.
  - Behave like you're in business for yourself: Consider how you, personally, can cut costs, serve the customer better, improve productivity, and innovate. Constantly think in terms of commercial success and how you and your group can add directly to the financial health of the organization.
  - Add value: Make sure you contribute more than you cost. It's your contribution that counts, not the hours, days, or years that you put into the company. You must compete for your position every day.

- Practice continuous improvement: Continuously improve yourself, your service, and your impact on the organization through small, ongoing improvements. Every day should be a pursuit of perfection.

MANAGEMENT OF INDUSTRIAL HYGIENE PROCESSES

- Management of industrial hygiene processes requires that the industrial hygienist:
  - Understand the organization's management style.
  - Define the organization's industrial hygiene needs.
  - Set priorities.
  - Communicate in the organization's language and style the mission, goals, objectives, plans, and tasks of the IH program.
  - Implement programs.
  - Measure effectiveness and progress.
  - Size resource needs.
  - Evaluate resources available by balancing resources and priorities, and considering the use of consultants.
  - Conduct audits and peer reviews.

Obstacles to a Successful Industrial Hygiene Program

- Obstacles to effective industrial hygiene programs include:
  - Lack of management acceptance and support.
  - Poor internal communications.
  - Differing attitudes.
  - Time limitations.
  - Understaffing.
  - Inability to keep up with change.

Benefits of an Industrial Hygiene Program

- A good industrial hygiene program results in a place of employment in which employees are protected from all known occupational health hazards in the workplace.

- Compensable injuries or illnesses are reduced, thus lowering insurance premiums and associated medical and recordkeeping costs.

- The improved working conditions usually increase productivity, reduce lost time accidents and illnesses, reduce absenteeism, and improve morale and labor relations.

- Anticipating and controlling potential occupational health hazards during the design phase of new projects reduces operating costs.

- OSHA and other government regulations concerning industrial hygiene are quickly assessed and implemented.

- Good programs provide a liaison with government regulatory agencies in dealing with occupational health issues.

- An effective industrial hygiene program offers an important source of health and safety information and training for employees.

- A good industrial hygiene program provides a liaison between the production areas and the medical department.

- Effective industrial hygiene programs also assist the medical department in determining whether a worker's medical condition is associated with a job-related health hazard or is an isolated event.

Glossary

A

Abrasive blasting: A process for cleaning surfaces by means of materials such as sand, alumina, or steel grit in a stream of high-pressure air.

Absorption: In air sampling, the capture of a gas or vapor accomplished by passing an air stream containing the gas or vapor through a liquid.

Acclimation: The process of becoming accustomed to new conditions, particularly heat. Also known as acclimatization.

Accuracy: The exactness of an observation obtained from an instrument or analytical technique with the true value. Often confused with precision.

ACGIH: See American Conference of Governmental Industrial Hygienists:

Acid pickling: Bath treatments to remove scale and other impurities from metal surfaces before plating or other surface treatment.

Action Level: A term used to express the level of toxicant that requires medical surveillance, air monitoring, and employee training, usually set at one-half the permissible exposure limit (PEL).

Activated charcoal: A gas or vapor adsorbent in solid sorbent sampling media. Charcoal is an amorphous form of carbon formed by burning wood, nutshells, and other carbonaceous materials. Activated charcoal is produced by heating charcoal with steam, forming a non-porous internal structure that gives it an extensive internal surface area.

Activity: Activity, usually expressed in terms of disintegrations per second, is often used as an abbreviation for radioactivity, and refers to the radiant power of a radioactive substance.

Acute: Health effects that show up a short length of time after exposure.

Administrative controls: Methods of controlling employee exposures by job rotation, work assignment, time periods away from the hazard, or training in specific work practices designed to reduce exposure.

Adsorption: The condensation of gases, liquids, or dissolved substances on the surfaces of solids such as activated charcoal or silica.

Aerodynamic diameter: The diameter of a unit density sphere with the same settling velocity as the particle in question.

Aerosol: Suspension of liquid or solid particles in air ranging in size from 0.01-100μm.

Air cleaner: A device designed to remove atmospheric airborne impurities, such as dusts, gases, vapors, fumes, and smokes.

Air horsepower: The theoretical horsepower required to drive a fan if there are no losses in the fan, and it is one hundred percent efficient.

Air monitoring: The sampling for and measurement of pollutants in the atmosphere.

Air mover: Any device that is capable of causing air to be moved from one space to another. These devices are used to exhaust, force, or draw gases through specific assemblies, such as exhaust ventilation systems.

Air-purifying respirator: Respirator that uses filters or sorbents to remove harmful substances from the air.

Air-supplied respirator: Respirator that provides a supply of breathable air from a clean source outside of the contaminated work area.

Airborne contaminants:

Aerosol: Suspension of liquid or solid particles in air ranging in size from 0.01-100μm.

Dust: Particulate material generated by a mechanical process ranging in size from 0.5-50μm.

Fiber: Particle with aspect ratio greater than 3:1 (NIOSH 5:1).

Fume: Solid particle aerosol formed by condensation from the vapor state, ranging in size from 0.001-0.2 μm.

Mist: Suspension of liquid particles in air formed by condensation from vapor or by some mechanical process ranging in size from 40-400 μm. Rain = 100μm.

Smoke: Aerosol formed from combustion of organic material ranging in size from 0.01-0.5 μm.

Airborne microorganisms: Biologically active contaminants suspended in air either as free-floating particles surrounded by a film or organic or inorganic material, or attached to the surface of other suspended particulates.

Aliphatic: An open-chain carbon compound. Usually refers to petroleum products derived from a paraffin base and having a straight or branched chain, and a saturated or unsaturated molecular structure. Methane and ethane are typical aliphatic hydrocarbons.

Alkali: A compound that has the ability to neutralize an acid and form a salt. Sodium hydroxide, known as caustic soda or lye, is an example. Sometimes referred to as a base, alkalis turn litmus paper blue.

Allergy: An abnormal response of a hypersensitive person to a chemical or physical stimuli.

Alpha emitter: A radioactive substance that gives off alpha particles.

Alpha particle: A small, positively charged particle made up of two neutrons and two protons and of very high velocity, generated by many radioactive materials, including uranium and radium.

Alveoli: Tiny air sacs in the lungs, located at the ends of bronchioles. Through the thin walls of the alveoli, blood takes in oxygen and gives up carbon dioxide in respiration.

Ambient noise: The all-encompassing noise associated with a given environment, usually a composite of sounds from many sources.

American Conference of Governmental Industrial Hygienists (ACGIH): An association whose membership is open to those engaged in the practice of industrial hygiene, occupational health and safety, or environmental health and safety.

American National Standards Institute (ANSI): A voluntary membership organization that develops consensus standards nationally for a wide variety of devices and procedures.

Amorphous: Noncrystalline.

Anechoic room: A room whose boundaries effectively absorb the entire sound incident therein, thereby affording essentially free field conditions.

Anemia: Deficiency in the hemoglobin and erythrocyte content of the blood. Term refers to a number of pathological states that may be attributed to a large variety of causes and appear in many different forms.

Anemometer: A device to measure air speed.

Anneal: Heating metals with subsequent cooling to temper or to soften them.

Anode: The positive electrode.

ANSI: See American National Standards Institute.

Anthrax: A highly virulent bacterial infection acquired from infected animals and animal products.

Anthropometry: The part of anthropology having to do with measurement of the human body to determine differences in individuals or groups of individuals.

APF: See assigned protection factor.

Aplastic anemia: A condition in which the bone marrow fails to produce an adequate number of red blood cells.

Aromatic: Applied to a group of hydrocarbons and their derivatives characterized by the presence of the benzene nucleus.

Asbestos: A hydrated magnesium silicate in fibrous form.

Asbestosis: A disease of the lungs caused by inhalation of fine airborne asbestos fibers.

Aspect ratio: Length to width ratio. Aspect ratio = L/W.

Asphyxia: Suffocation from lack of oxygen. A substance such as carbon monoxide that combines with hemoglobin to reduce the blood's capacity to transport oxygen produces chemical asphyxia. Simple asphyxia is the result of exposure to a substance, such as methane, that displaces oxygen.

Asphyxiant: A gas whose primary or most acute health effect is asphyxiation. There are two classes of asphyxiants: simple asphyxiants, such as nitrogen or methane, which act by replacing oxygen; and chemical asphyxiants, such as carbon monoxide, which cause asphyxiation by preventing oxygen uptake at the cellular level.

Assigned protection factor (APF): The level of respiratory protection expected from a respirator that is properly functioning, has been properly fitted, and is worn by a worker trained in its use. APFs can be used to help provide an estimate of the maximum concentrations of a contaminant in which a particular respirator can be used.

Asthma: Constriction of the bronchial tubes in response to irritation, allergy, or other stimulus.

Atmospheric pressure: The pressure exerted in all directions by the atmosphere. At sea level, mean atmospheric pressure is 29.92 in. Hg, 14.7 psi, or 407 in. wg.

Atomic weight: The atomic weight is approximately the sum of the number of protons and neutrons found in the nucleus of an atom. This sum is also called the mass number. The atomic weight of oxygen is approximately 16, with most oxygen atoms containing 8 neutrons and 8 protons.

Attenuate: To reduce in amount. Used to refer to noise or ionizing radiation.

Audiogram: A record of hearing loss or hearing level measured at several different frequencies, usually 500-6000 hertz. The audiogram may be presented graphically or numerically.

Audiometer: A signal generator or instrument for objectively measuring the sensitivity of hearing. Pure-tone audiometers are standard instruments for industrial audiometric testing.

Autoclave: An apparatus using pressurized steam for sterilization.

Avogadro's Number: The number of molecules in a mole of any substance (6.02217×10^3). Named after Italian physicist Amedeo Avogadro (1776-1856). At 0° C and 29.92 in. Hg, one mole occupies 22.4 liters of volume.

Axial Flow Fan: A propeller type fan useful for moving large volumes of air against little resistance.

B

Bacillus: A rod-shaped bacterium.

Background noise: Noise coming from sources other than the particular noise sources being monitored.

Background radiation: The radiation coming from sources other than the radioactive material to be measured. This background is primarily due to cosmic rays that constantly bombard the earth from outer space.

Bacteria: Microscopic organisms living in soil, water, organic matter, or the bodies of plants and animals. They are characterized by the lack of a distinct nucleus and the inability to photosynthesize.

Bag house: The housing containing bag filters for recovery of fumes of arsenic, lead, sulfur, etc., from the flues of smelters.

Basal metabolism: A measure of the amount of energy required by the body at rest.

Bauxite: Impure mixture of aluminum oxides and hydroxides. The principal source of aluminum.

Beam divergence: Angle of beam spread measured in mrad (mrad = 3.4 min. of arc). Usually used in laser applications.

Becquerel (Bq): One disintegration per second. A measure of the rate of radioactive disintegration. There are 37 billion Bqs per curie.

BEIs: See biological exposure indices.

Belding-Hatch Index: Also referred to as Heat Stress Index. Estimate of the body heat stress of a standard man for various degrees of activity. Also relates to sweating capacity.

Benzene: A major organic intermediate and solvent derived from coal or petroleum. The simplest member of the aromatic series of hydrocarbons.

Berylliosis: Chronic beryllium intoxication.

Beta decay: The process whereby some radioactive emitters give off a beta particle. Also called beta disintegration.

Beta particle: Beta radiation. A small electrically charged particle thrown off by many radioactive materials, identical to the electron. Beta particles emerge from radioactive material at high speeds.

Biohazard: An abbreviation of biological hazard. Organisms or products of organisms that present a risk to humans.

Biological exposure indices (BEIs): Advisory biological limit values adopted by the ACGIH for some substances. Indices are based on urine, blood, or expired air samples. A BEI may be a value for the substance itself or it may refer to a level of a metabolite. BEIs represent the value of the biological determinant that is most likely to be the value of that determinant obtained from a worker exposed at the eight-hour TLV-TWA for the substance in question.

Biological half-life: The time required to reduce the amount of an exogenous substance in the body by half.

Biomechanics: The study of the human body as a system operating under two sets of laws: the laws of Newtonian mechanics and the biological laws of life.

Bloodborne pathogen: A pathogen where there is risk of exposure to blood or other potentially infectious material. Hepatitis B and HIV are the most often discussed pathogens.

Body burden: The amount of material in the body at a given time.

Bq: See becquerel.

Brake horsepower: The horsepower required to drive a unit. It includes the energy losses in the unit and can only be determined by actual testing. It does not include drive losses between the motor and the unit.

Braze: To solder with any relatively infusible alloy. A brazing furnace is a high temperature furnace used for heating metals to be joined by brazing.

Breathing zone: A globe-shaped area with a two-foot radius surrounding the head.

Breathing zone sample: An air sample collected in the breathing zone of workers to assess their exposure to airborne contaminants.

Bremsstrahlung: Secondary x-radiation produced when a beta particle is slowed down or stopped by a high-density surface.

Bronchial tubes: Branches or subdivisions of the trachea. A bronchiole is a branch of a bronchus.

Bronchitis: Inflammation of the bronchi or bronchial tubes.

Brownian movement: The irregular movement of particles suspended in a fluid as a result of bombardment by atoms and molecules.

Brucella: A genus of short, rod-shaped to coccoid, encapsulated, gram-negative, parasitic, pathogenic bacteria.

Brucellosis: A group of diseases caused by an organism of the *brucella* genus. Also called undulant fever. One source is unpasteurized milk from cows suffering from Bang's disease.

Byssinosis: Disease resulting from prolonged exposure to heavy air concentrations of cotton or flax dust.

C

Calorimeter: A device for measuring the total amount of energy absorbed from a source of electromagnetic radiation.

Cancer: A cellular tumor formed by mutated cells.

Capture velocity: Air velocity at any point in front of an exhaust hood necessary to overcome opposing air currents and to capture the contaminated air by causing it to flow into the hood.

Carbon monoxide: A colorless, odorless gas produced through incomplete combustion processes.

Carboxyhemoglobin: A reversible combination of carbon monoxide with hemoglobin.

Carcinogenic: Cancer causing.

Carcinoma: Malignant or cancerous tumor.

Carpal tunnel: A passage in the wrist through which the median nerve and many tendons pass to the hand from the forearm.

Cataract: Opacity in the lens of the eye that may obscure vision.

Caustic: Something that strongly irritates, burns, corrodes, or destroys living tissue.

Ceiling limit (C): An airborne concentration of a toxic substance in the work environment that must never be exceeded.

Centrifugal fan: Wheel-type fan useful where static pressure is medium to high.

CET: See corrected effective temperature.

CFR: See Code of Federal Regulations.

Chemical cartridge: The type of absorption unit used with a respirator for removal of low concentrations of specific vapors and gases.

Chloracne: Acne originating from exposure to chlorinated naphthalenes and polyphenyls.

Chromatograph: An instrument that separates and analyzes mixtures of chemical substances.

Chronic: Long latency period. Persistent and prolonged.

Ci: See Curie.

Clostridium botulinum: Human pathogen bacteria that produce a neurotoxin, botulin, that causes botulism.

Cochlea: The auditory part of the internal ear, shaped like a snail shell. It contains the basilar membrane on which the end organs of the auditory nerve are distributed.

Code of Federal Regulations (CFR): The rules promulgated and enforced under US law, and published in the *Federal Register*, are incorporated in this code.

Coefficient of entry: The ratio of actual flow to theoretical flow.

Colorimetry: A term used for all chemical analysis involving reactions in which a color is developed when a particular contaminant is present in the sample and reacts with the collection medium. The resultant color intensity or stain length is measured to determine the actual concentration.

Combustible liquids: Liquids are considered combustible if they have a flash point above 100° F.

Communicable: A disease whose causative agent is readily transmitted from one person to another.

Compound: A substance composed of two or more elements joined by chemical bonds.

Concentration: The amount of a given substance in a stated unit of measure, expressed as percentage weight by volume, weight per unit volume, etc.

Conductive hearing loss: Type of hearing loss usually caused by a disorder effecting the middle or external ear.

Conjunctivitis: Inflammation of the conjunctiva of the eye.

Contact dermatitis: Dermatitis caused by contact with a substance.

Convection: The motions in fluids resulting from a difference in density and the action of gravity.

Coolant: A transfer agent used in metal machining to remove heat from the cutting tool and metal interface. A coolant is sometimes referred to as a cutting fluid.

Cornea: Transparent membrane covering the anterior portion of the eye.

Corrected effective temperature (CET): An index of thermal stress similar to the effective temperature index, except that the globe temperature is used instead of the dry bulb temperature.

Corrosive: A substance that causes destructive and permanent changes in human skin tissue at the site of contact.

Coulometer: A chemical analysis instrument that determines the amount of a substance released in electrolysis by measurement of the quantity of electricity used. The number of electrons transferred in terms of coulombs is an indication of the contaminant concentration.

Cristobalite: A crystalline form of free silica.

CTD: See cumulative trauma disorder.

Cubic centimeter (cm^3): A volumetric measurement that is equal to one milliliter (ml).

Cubic meter (m^3): A measure of volume in the metric system.

Cumulative trauma disorder (CTD): A disorder of a musculoskeletal or nervous system component caused by repeated and or forceful movements of said musculoskeletal system.

Curie (Ci): A measure of the rate at which radioactive materials decay. Named after French chemists Marie Curie (1867-1934) and Pierre Curie (1859-1906).

Cyclone device: A dust-collecting instrument that has the ability to separate particles by size.

D

Dampers: Adjustable sources of airflow resistance used to distribute airflow in ventilation systems.

dBA: Sound level in decibels measured on the A scale of a sound level meter.

dBC: Sound level in decibels measured on the C scale of a sound level meter.

Decibel: A unit for expressing the ratio of two amounts of electric or acoustic power equal to ten times the common logarithm of this ratio or 20 times the common logarithm of acoustic voltage or current ratio. A unit for expressing the relative intensity of sounds on a scale from zero for the average least perceptible sound to about 130 for the average pain level.

Decontaminate: To rid of contamination and to make safe by eliminating poisonous, radioactive, or other harmful substances.

Density: The ratio of mass to volume. $D = M/V$

Dermatitis: Inflammation of the skin from any cause.

Dermatosis: A broader term than dermatitis, it includes any skin abnormality including folliculitis, acne, pigment changes, etc.

Dioctyl phthalate (DOP): A powdered chemical that can be aerosolized at a uniform rate. Typically used in air-purifying respirator testing methods.

Direct current (dc): An electric current which only flows in one direction.

Direct reading instruments: Devices that provide an immediate indication of the concentration of aerosols, gases, and vapors by means of a color change in colorimetric devices, or a register on a meter or instrument.

Disintegration: A decay process that results in the release of energy in the form of radiation.

Dispersion: The mixing or dilution of a contaminant in the atmosphere.

DOP: See dioctyl phthalate.

Dose: A term used to express an amount of exposure to a chemical substance.

Dose-response relationship: Correlation between the amount of exposure to an agent or toxic chemical and the resulting effect on the body.

Dosimeter: A device used to determine the full-shift exposure a person has received to a physical hazard.

Droplet: A liquid particle suspended in a gas.

Dry bulb temperature: An ordinary thermometer. The temperature of a gas or mixture of gases indicated on a thermometer after correction for radiation.

Duct: A conduit used for moving air at low pressures.

Duct velocity: Air velocity through the duct cross section.

Dust collector: An air cleaning device used to remove heavy particulate loadings in exhaust systems before discharge to the outdoors.

Dusts: Solid particles generated by a mechanical process, such as handling, crushing, and grinding of organic or inorganic materials. Particle size ranges from 0.5-50μm.

Dyspnea: Shortness of breath, difficult or labored breathing.

E

ECT: See effective chill temperature.

Effective chill temperature (ECT): A temperature index used to consider heat loss from skin exposed to the combined effects of cold temperatures and air speed. Also known as wind chill factor and equivalent chill temperature.

Effective temperature (ET): An arbitrary index that combines into a single value the effects of temperature, humidity, and air movement on the sensation of warmth and cold on the human body.

Electrostatic precipitator: An instrument that removes particles from an air stream by charging the particles and collecting them on a suitable surface. Also known as electrical precipitator.

Electromagnetic spectrum: The entire range of wavelengths or frequencies of electromagnetic radiation extending from gamma rays to the longest radio waves and visible light.

Electron volt (eV): A unit of energy equal to the energy gained by an electron in passing from a point of low potential to a point one volt higher in potential: 1.6×10^{-19} joule.

Electroplate: To cover with a metal coating through electrolysis.

Elutriator: A device used to separate particles according to mass and aerodynamic size by maintaining a laminar flow system at a rate that permits the heavier particles to settle more rapidly while smaller particles are kept airborne by resistance forces of the flowing air.

Engineering controls: Methods of controlling worker exposures by modifying the source or reducing the quantity of contaminants released into the work environment.

Entry loss: Loss in pressure caused by air flowing into a duct or hood.

Epicondylitis: Inflammation of certain bony prominences in the area of the elbow. Also known as tennis elbow.

Erythema: Reddening of the skin.

ET: See effective temperature.

Evaporation rate: The ratio of the time required to evaporate a measured volume of a liquid to the time required to evaporate the same volume of a reference liquid under test conditions.

Excursion limits: Short-term exposures should exceed three times the threshold limit value-time-weighted-average (TLV-TWA) for no more than a total of 30 minutes during a work day and under no circumstances should they exceed five times the TLV-TWA.

Exhalation valve: A valve that permits exhaled air to leave a respirator and prevents outside air from entering through the valve.

Exhaust ventilation: A mechanical device used to remove air from any space.

Exposure: Contact, usually through inhalation, absorption, or ingestion, with a chemical, biological, or physical hazard.

F

Face piece: The component of a respirator that covers the wearer's nose, mouth, and possibly eyes. The face piece provides a gas or dust-tight seal with the face and is held in place with headbands.

Face velocity: Average air velocity into an exhaust system measured at the opening into the hood or booth.

Fan laws: Theories and equations that describe the relationships between fan volume, pressure, brake horsepower, size, and rotating fan speed.

Fan static pressure: The pressure added to the system by the fan. It equals the sum of pressure losses in the system less the velocity pressure in the air at the fan inlet.

Far field: The distance from the noise source where the sound-pressure level decreases 6 dBA for each doubling of distance.

Fiber: Particle with an aspect ratio greater than 3:1 (NIOSH 5:1).

Fibrosis: A lung condition identified by an increase of interstitial fibrous tissue (scars).

FID: See flame ionization detector.

Film badge: A dosimeter worn by workers to measure their exposure to ionizing radiation. It is a small piece of photographic film darkened by nuclear radiation. The film is inspected at a laboratory and quantifies radiation exposure.

Flame ionization detector (FID): A direct reading instrument that ionizes gases and vapors with an oxyhydrogen flame and measures the differing electrical currents thus generated.

Flammable liquid: Any liquid having a flash point below 100° F.

Flash point: The lowest temperature at which a liquid gives off enough vapor to form an ignitable mixture with air and produce a flame when a source of ignition is present.

Flow meter: A device for measuring the flow rate of a fluid or gas.

Flux: A substance used to clean surfaces and promote fusion in soldering.

Foot-candle: A unit of illumination. It is the illumination on a surface one-foot from, and perpendicular to, a uniform point source of one candle.

Free field: A field free of boundaries, usually used when evaluating noise levels.

Frequency: The number of complete oscillations per second of energy (sound or electromagnetic radiation) in the form of waves. Measured in hertz (Hz).

Friction loss: The pressure loss caused by friction.

Fume: Airborne particulates created by the condensation of solid particles from the gaseous state. Usually generated from a combustion process or a melting process, such as welding. Particle size ranges from 0.001-0.2 μm.

Fundus: The interior surface of a hollow organ, such as the retina of the eye.

G

Galvanizing: The process of adding a protective coat of zinc to metal.

Gamma rays: High energy x-rays.

Gas chromatography: A detection technique that separates a gaseous mixture by passing it through a column, enabling the components to be released at various times depending on their molecular structure. Used as an analytical tool for air sampling gases and vapors.

Geiger-Muller counter: A gas-filled electrical instrument that indicates the presence of an atomic particle or ray by detecting the ions produced.

General ventilation: A ventilation system using natural or mechanically generated make-up air to mix and dilute contaminants in the workplace.

Globe thermometer: A thermometer set in the center of a black metal sphere to measure radiant heat.

Grab sample: A sample taken within a short duration to quantify or identify air contaminants.

Gram (g): A metric unit of weight.

Granuloma: A mass or nodule of inflamed tissue.

Gravimetric: Measurement by weight. Usually used for quantifying dusts by weighing an air sampling filter before use and then weighing the filter after exposure.

Gravity: The force of attraction exerted on objects by the Earth's gravitational field, equal to 32.2 ft/sec^2 or 98 cm/sec^2.

Gray (Gy): A unit of absorbed radiation dose equal to one joule of absorbed energy per kilogram of matter.

Gy: See Gray.

H

Half-life, radioactive: For a single radioactive decay process, the time required for the activity to decrease to half its value by that process.

Half-value layer (HVL): The thickness of a substance necessary to reduce the intensity of a beam of gamma or x-rays to half its original value. Also known as half-thickness.

Halogenated hydrocarbon: A chemical material consisting of carbon and one or more halogens (chlorine, fluorine, astatine, bromine, or iodine).

Halogens: Collective name given to five non-metallic elements: astatine, bromine, chlorine, fluorine, and iodine.

Hazardous material: Any substance that has the capacity to produce adverse effects on the health and safety of humans.

Heat cramps: Muscle spasms resulting from exposure to excessive heat.

Heat exhaustion: A condition usually caused by loss of water due to exposure to excessive heat. Symptoms include headache, nausea, and fainting.

Heat rash: A rash caused by sweating and inadequate hygiene practices.

Heat stress index (HSI): Also known as the Belding-Hatch heat stress index. The index combines environmental heat and metabolic heat in determining the body's requirements for evaporation of sweat.

Heat stroke: A serious disorder resulting from exposure to excessive heat. Caused by sweat suppression and increased storage of body heat. Symptoms include hot, dry skin, high temperature, mental confusion, convulsions, and coma.

Heat syncope: A heat-related disorder characterized by blurred vision and brief fainting spells. Caused by pooling of blood in the legs or skin during prolonged static postures in a hot environment.

Hematuria: Blood in urine.

Hemoglobin: The protein blood that carries oxygen. Gives blood its characteristic red color when oxygenated.

HEPA filter: See high efficiency particulate air (HEPA) filter

Hepatitis B: A bloodborne virus causing hepatitis.

Hepatotoxin: A chemical that induces liver damage.

Hertz (Hz): The frequency measured in cycles per second. 1 cps = 1 Hz.

High efficiency particulate air (HEPA) filter: A disposable, extended-medium, dry-type filter with a particle removal efficiency of not less than 99.97 percent for 0.3 micron particles.

High frequency loss: A hearing deficit starting at 2000 Hz.

HIV: See human immunodeficiency virus.

Hood: The enclosure part of a local exhaust system. Also, a device that completely covers the wearer's head, neck, and portions of the shoulders.

Hood entry loss: The pressure loss from turbulence and friction as air enters a ventilation system.

Hood slot: A hood consisting of a narrow slot leading into a plenum chamber under suction to distribute air velocity along the length of the slot.

Hood static pressure: The suction or static pressures in a duct near a hood. It represents the suction that is available to draw air into the hood.

Horsepower: A unit of power, equivalent to 33,00 foot pounds per minute.

Host: A plant or animal harboring another as a parasite or as an infectious agent.

HSI: See heat stress index.

Human immunodeficiency virus (HIV): The human immunodeficiency virus causes acquired immunodeficiency syndrome (AIDS).

HVL: See half-value layer.

Hydrocarbons: Organic compounds composed of carbon and hydrogen. Major natural sources of hydrocarbons are petroleum, natural gas, and coal.

Hygroscopic: Readily absorbing moisture.

Hypothermia: Reduced body temperature, a symptom of cold stress.

Hz: See Hertz.

I

IDLH: See Immediately Dangerous to Life and Health.

Illumination: The amount of light falling on a surface, measured in footcandles.

Immediately Dangerous to Life and Health (IDLH): A maximum concentration of a substance from which one could escape within 30 minutes without any escape-impairing symptoms or any irreversible health effects.

Impaction: Forcibly lodging particles into matter.

Impingement: The process of collecting particles by pulling contaminated air through a device filled with water or reagent (particles remain in the liquid).

Inches of mercury (in. Hg): A unit used in measuring pressures. One inch of mercury column equals 0.491 psi of pressure.

Inches of water (in. H_2O): A unit used in measuring pressures. One inch of water column equals 0.036 psi of pressure.

Inert: A substance with inactive properties.

Infection: A state in the body produced upon exposure to an infective agent, whether the agent is a microorganism, bacteria, parasite, etc.

Infectious: Capable of causing infection.

Inflammation: A local response to a cellular injury earmarked by redness and swelling.

Infrared: Wavelengths of the electromagnetic spectrum longer than those of visible light and shorter than radio frequency.

Ingestion: Refers to a method of entry for hazardous chemicals by taking in as for digestion.

Inhalation valve: A component of a respirator that allows contaminated air to pass through to a filter or purifier prior to inhalation. This device also prevents exhaled air from leaving the respirator by closing under positive pressure.

Inorganic: Refers to compounds and substances that do not contain carbon.

Intensity: The amount of light a source generates in a given direction, measured in candela.

Intoxication: An abnormal state associated with poisoning.

Intraperitoneal: Existing or entering through the peritoneum.

Inverse square law: A given physical quantity varies with the distance from the source inversely as the square of the distance. Applies to radiation and illumination.

Inversion: An atmospheric condition where a layer of warm air traps a layer of cool air below it so that the bottom layer cannot rise.

Ion: An electrically charged atom. An atom that has lost one or more electrons is considered a positive ion, and an atom that has gained one or more electrons is considered a negative ion.

Ion chamber: An instrument used to measure ionizing radiation. Also known as ionization chamber.

Ionizing radiation: A range in the electromagnetic spectrum that interacts with matter to produce ions. The five types of ionizing radiation are alpha, beta, x-ray, gamma, and neutron.

Irritant: A substance or material that produces redness, swelling, or pain when in contact with skin, eyes, nose, or respiratory system.

J

Jaundice: Yellowish pigmentation of the skin, eyes, and body fluids caused by the deposition of bile pigments.

Joule: Unit of energy used in describing a single pulsed output of a laser.

Joules per square centimeter (Joule/cm^2): Unit of energy density used in measuring the amount of energy per area of absorbing surface or per area of a laser beam.

K

Kelvin: A temperature scale, also called absolute temperature, where the temperature is measured on the average kinetic energy per molecule of a perfect gas. Named for British physicist William Thomson, Lord Kelvin (1824-1907).

Kilogram (kg): A unit of weight in the metric system equal to 2.2 pounds.

L

Laminar flow: Streamline flow of air within a solid boundary. The opposite of laminar flow would be turbulent flow.

Laser: Light amplification by stimulated emission of radiation.

Latency period: The time period between exposure and the manifestation of disease or damage.

LC_{50}: Lethal concentration is the concentration at which fifty percent of test animals die.

LCL: See Lower Confidence Limit.

LD_{50}: Lethal dose is the dose at which fifty percent of test animals die.

LEL: See Lower Explosive Limit.

Lethal Concentration: See LC_{50}.

Lethal Dose: See LD_{50}.

Leukemia: An acute or chronic disease characterized by abnormal production of leukocytes or white blood cells.

Liquid: A state of matter in which the substance is a formless fluid that follows the principles of gravity.

Liter (l): A measure in the metric system of volume. One liter equals 1.057 quarts.

Local exhaust ventilation: A ventilation system that captures and removes contaminants at the point of generation before escaping into the workplace.

Lower confidence limit (LCL): The likelihood that the true value of the sample quantity is lower than that obtained.

Lower explosive limit (LEL): The lowest vapor concentration boundaries within which the vapor-air mixture will propagate flame if it is ignited.

Lumen: A unit of luminous flux equal to the light emitted in a unit solid angle by a uniform point source of one candle intensity.

Lux: A unit equal to one lumen per square meter. Also a unit of illumination equal to 10 foot-candles.

Lyme disease: A disease transmitted to humans by the deer tick.

M

Macrophage: A phagocytic cell that engulfs foreign matter in the body. Also called a histiocyte.

Makeup air: Clean, tempered outdoor air supplied to a workspace to replace air removed by exhaust ventilation.

Manometer: A device used for measuring the pressure of gases and liquids.

Maximum use concentration (MUC): The product of the protection factor of respiratory protection equipment and the permissible exposure limit (PEL).

Membrane filter: A filter usually made from cellulose or polyethylene with various pore diameters. Used in air sampling for the collection of a wide range of metals and minerals.

Mesothelioma: Cancer of the membranes that line the chest and abdomen, almost exclusively associated with asbestos exposure.

Metal fume fever: Flu-like condition caused by inhaling metal fumes. Usually associated with zinc and manganese.

Methemoglobin: A soluble, brown, crystalline basic blood pigment that differs from hemoglobin in that it contains ferric iron and is unable to combine reversibly with molecular oxygen.

Microcurie: One millionth of a curie.

Micron (μ) or micrometer (μm): A unit of length in the metric system equal to 10^{-4} cm, approximately 1/25,000 of an inch.

Microorganism: Any minute organism, including bacteria, viruses, molds, microbes and others.

Milligram (mg): A unit of weight in the metric system. One thousand milligrams equals one gram.

Milligrams per cubic meter (mg/m^3): A unit used to measure air concentrations of dusts, gases, mists, and fumes.

Milliliter (ml): A unit of the metric system that measures volume. One milliliter equals one cubic centimeter.

Millimeter of mercury (mm Hg): The unit of pressure equal to the pressure exerted by a column of liquid mercury one millimeter high at standard temperature.

Mists: Suspended liquid particles in air generated by condensation from the gaseous to the liquid state or by some mechanical process atomizing a liquid. Particles range in size from 40-400 μm. Rain = 100μm.

Monaural hearing: Hearing with one ear only.

MUC: See maximum use concentration.

Mucous membranes: Lining of the hollow organs of the body, the nose, mouth, stomach, intestines, bronchial tubes, and urinary tract.

Mutagen: A substance or material that causes change in the genetic material of a cell.

N

Nanometer: A unit of length equal to 10^{-7} centimeters.

Narcosis: Stupor or unconsciousness produced by chemical substances.

Near field: Refers to a field in the immediate vicinity of a noise source where the sound-pressure level does not follow the inverse square law.

Nephrotoxin: A substance or material poisonous to the kidney.

Neurotoxin: A substance or chemical toxic to the nervous system.

NIHL: See noise induced hearing loss.

Noise induced hearing loss (NIHL): A slow, progressive loss of hearing related to high noise levels over a long period of time.

Non-ionizing radiation: Electromagnetic radiation that does not cause ionization. Includes ultraviolet, laser, infrared, microwave, and radio frequency radiation.

Nuisance dust: Dust that does not produce an adverse health effect. Considered an irritant, thus earning the name "nuisance dust."

O

Octave: The interval between two frequencies (as in an electromagnetic spectrum) having a ratio of 2 to 1.

Octave band: An arbitrary spread of frequencies. The top frequency is twice the bottom frequency. Usually referred to by its center frequency.

OD: See optical density.

Odor threshold: The minimum concentration of a substance at which a majority of test subjects can detect and identify the characteristic odor of a chemical.

Olefin: An unsaturated hydrocarbon obtained from petroleum and natural gas. Usually identified by the formula C_nH_{2n}.

Optical density (OD): A logarithmic expression of the attenuation afforded by a filter.

Organic: Chemicals that contain carbon.

Orifice: A critical or limiting orifice is used to control the flow rate of a gas in rotameters and other air sampling equipment.

OSHA 200 log: OSHA requires this form be used to record employee injuries and illnesses.

Otosclerosis: A growth of spongy bone in the inner ear that causes progressive deafness.

Overexposure: Exposure outside or beyond the recommended or regulatory limits.

P

Paraffin: Straight or branched chain hydrocarbon components of crude oil and natural gas whose molecules are saturated and therefore very stable. Usually identified by the formula C_nH_{2n+2}.

Particle: A small mass of a solid or liquid.

Particulate: A particle.

Particulates: A suspension of fine solid or liquid particles in air, including dusts, mists, fog, fume, smoke or sprays.

Pathogen: A microorganism capable of causing disease.

PEL: See permissible exposure limit.

Peripheral neuropathy: Destruction of the peripheral nervous system, including the nerves to the hands, arms, legs, and feet.

Permissible exposure limit (PEL): An exposure limit published and enforced by OSHA. The maximum time-weighted average exposure allowed by OSHA for an eight-hour workday during a normal 40-hour work week.

Personal protective equipment (PPE): Equipment worn by workers to protect against hazards, including gloves, clothing, glasses, boots, etc.

Phenol: Intermediate chemicals and bases for plastics, explosives, antiseptics, and pharmaceuticals. Sometimes referred to as carbolic acid.

Photoelectric effect: Occurs when an electron is released from an atom by a light ray or gamma ray.

Photoionization detector (PID): A direct reading monitoring instrument that operates by detecting and distinguishing between ions of vapors and gases following ionization by the instrument's ultraviolet light source.

PID: See photoionization detector.

Pitot tube: A device used for measuring static pressure within ventilation ducts.

Plenum: Pressure equalizing chamber.

Pneumoconiosis: A lung disease caused by inhalation of various types of dusts.

Polar solvents: Oxygen-containing chemicals with high dielectric properties.

PPE: See personal protective equipment.

Precision: The degree of exactness of repeated measurements.

Presbycusis: Age-induced hearing loss.

Pressure: Force applied over a surface, measured as force per unit area.

Pressure drop: The difference in static pressure measured at two locations in a ventilation system.

Pressure loss: Energy lost in a duct system through friction or turbulence.

Proton: A unit of matter having a positive charge and a mass of one.

Psittacosis: Also known as parrot fever. An infectious disease of birds to which poultry handlers and other workers exposed to dried bird feces are at risk. Caused by *chlamydia psittaci.*

Psychrometer: A device consisting of wet and dry bulb thermometers for measuring relative humidity. A psychrometric chart is a graphical representation of the thermodynamic properties of moist air.

Pulsed laser: A class of laser that operates in the pulsed mode, with repetitive short duration emissions.

Push-pull ventilation hood: A hood that exhausts air on one side of an open-surface tank. Air is forced across the tank from a slotted hood that pushes it toward the exhaust side of the tank.

Q

Q fever: A disease caused by a rickettsial organism that infects meat and livestock handlers.

Q-switched laser: A pulsed laser capable of extremely high peak powers for very short durations.

Quality factor: A linear energy transfer-dependent factor by which absorbed radiation doses are to be multiplied to obtain the dose equivalent.

Quartz: A naturally occurring form of free silica found in sandstone and sand.

R

Rad: Roentgen absorbed or radiation absorbed dose.

Radiant temperature: The temperature generated by a material absorbing radiant energy.

Rash: Reddening or blotching of the skin.

Raynaud's syndrome: An abnormal constriction of blood vessels in the fingers when exposed to cold temperature. Caused by vibrating hand tools.

Recommended exposure limit (REL): An exposure limit developed by NIOSH based on research and toxicological data.

REL: See recommended exposure limit.

Relative biological effectiveness: The relative effectiveness of the same absorbed dose of two ionizing radiations in producing a measurable biological response.

Reliability: The ability to perform consistently over time (such as air sampling equipment).

rem: See Roentgen equivalent man.

Reticule: A scale, grid, or other pattern located in the focus of the eyepiece of a microscope. Also known as a reticle.

Rickettsia: Rod-shaped microorganisms that live within cells. Usually transmitted via arthropods.

Roentgen: A unit of radioactive dose. Named after German physicist Wilhelm Konrad Roentgen (1845-1923).

Roentgen equivalent man (rem): A radiation dose unit that equals the dose in rads multiplied by a quality factor or relative biological effect for a specific radiation.

Rotameter: A small tapered tube with a solid ball (float) inside used to measure flow rate of air sampling equipment.

S

SAE: See sampling and analytical error.

Salmonella: A gram-negative, rod-shaped, disease-causing bacteria.

Sampling and analytical error (SAE): An analytical or sampling error that may cause a sample result to deviate from the true result.

Scintillation counter: An instrument for counting particles with light (scintillation). Used for measuring ionizing radiation.

Sealed source: A radioactive isotope or material sealed in a container to prevent accidental release of radioactivity.

Sensitivity: The minimum amount of an air contaminant repeatedly detected.

Sensitizer: An allergen. A material capable of producing an allergic response.

Sensorineural: Hearing loss caused by inner ear damage to the sensory or nervous system.

Shaver's disease: Aluminum-induced pneumoconiosis.

Shields: A variety of materials used to block or stop the transmission of radiation.

Siderosis: A lung disease associated with iron fume exposure.

Sievert (Sv): A unit of absorbed radiation dose in Gray (Gy) times the quality factor of the radiation in comparison to gamma-radiation. 1 Sv = 100 rem.

Silica gel: An adsorbent material used in air sampling for inorganic chemicals, such as acid mists.

Silicosis: A lung disease caused by the deposition of silica dust in the lungs.

Slot velocity: The flow rate through a slot-type hood.

Smoke: An air suspension (aerosol) of particles generated by combustion or sublimation of organic material. Particles range in size from 0.01-0.5 μm.

Soap bubble meter: An upturned burette used to calibrate air sampling pumps.

Solder: A material used for joining metal surfaces together by forming a joint (filling a junction).

Solid-state laser: A laser that uses a solid crystal, such as ruby or glass, as its excitation medium.

Solvent: A substance that dissolves another substance.

Sorbent: A material used to collect gases and vapors during air sampling.

Sound pressure level (SPL): The level (in decibels) of a sound is twenty times the logarithm to the base 10 of the ratio of the pressure of said sound to the reference pressure.

Specific gravity: The ratio of the density of a substance to the density of some substance taken as a standard when both densities are obtained by weighing in air.

Specificity: The exactness of a detection method to accurately detect a single contaminant in the presence of other contaminants.

Spectrophotometer: A direct reading instrument used for comparing the intensities of electromagnetic wavelengths produced by absorption of ultraviolet, visible, or infrared radiation from a gas or vapor.

SPL: See sound pressure level.

Standard air: Air at standard temperature and pressure.

Stannosis: A pneumoconiotic disease of the lungs caused by the inhalation of tin-containing dusts.

Static pressure: The potential pressure exerted in all directions by air within a duct system. It is the tendency to burst or collapse a pipe, and is expressed in inches (in.) of water gauge (wg).

STEL: See Threshold Limit Value-Short Term Exposure Limit.

Student's t-test: Named for William Gossett, an English statistician who used the pseudonym "Student." Gossett found that the shape of a distribution depended on the sample size. Student's t-test is used for small sample sizes.

Symptom: The subjective feelings of a patient.

Synergistic: The capacity to act in synergism. When two substances interact to produce a sum greater than the individual substances, or when two chemicals together produce more dangerous health effects than one chemical alone, they are deemed synergistic.

Systemic: Affecting all organs of the body rather than remaining localized and producing a local effect.

T

Temper: To soften metals by heating at a lower temperature, or to harden metals by heating and cooling in oil. Also, to anneal glass by gradually heating and cooling.

Temporary Threshold Shift (TTS): Temporary hearing loss due to noise exposure. May be partially or completely recovered when exposure ends.

Tendon: A tough cord or band of dense white fibrous connective tissue that unites a muscle with some other part (as a bone) and transmits the force that the muscle exerts.

Tendonitis: Inflammation of a tendon.

Tenosynovitis: Inflammation of the connective tissue sheath of a tendon.

Teratogen: A substance that may cause physical defects in the developing embryo or fetus when a pregnant female is exposed to that substance.

Terminal velocity: The fall rate of a particle through a fluid caused by external forces, including gravity.

Threshold Limit Value-Ceiling (TLV-C): The concentration of a substance to which exposure should not be exceeded during any part of the working exposure. In conventional industrial hygiene practice, if instantaneous monitoring is not feasible, the TLV-C applies to a 15-minute period, except for those substances that may cause immediate irritation when exposures are short.

Threshold Limit Value-Short Term Exposure Limit (TLV-STEL): The concentration of a substance to which workers can be exposed continuously for a short period of time without suffering from irritation, chronic or irreversible tissue damage, or narcosis of sufficient degree to increase the likelihood of accidental injury, impairment of self-rescue, or reduction in work efficiency, provided that the daily TLV-TWA is not exceeded. The TLV-STEL is not a separate independent exposure limit. It supplements the time-weighted average (TWA) limit where there are recognized acute effects from a substance whose toxic effects are primarily of a chronic nature. STELs are recommended only where toxic effects have been reported from high short-term exposures in either humans or animals. A STEL is defined as a 15-minute TWA exposure that should not be exceeded at any time during a workday even if the eight-hour TWA is within the TLV-TWA. Exposures above the TLV-TWA up to the STEL should not be longer than 15 minutes and should not occur more than four times per day. There should be at least 60 minutes between successive exposures in this range. An averaging period other than 15 minutes may be recommended when this is warranted by observed biological effects.

Threshold Limit Value-Time-Weighted Average (TLV-TWA): The time-weighted average concentration for a normal eight-hour workday and a 40-hour workweek, to which nearly all workers may be repeatedly exposed, day after day, without adverse effect.

Time-weighted-average (TWA): Refers to concentrations of airborne contaminants weighted for time (usually eight hours).

Tinnitus: The perception of ringing in the ears.

TLV-C: See Threshold Limit Value-Ceiling.

TLV-STEL: See Threshold Limit Value-Short Term Exposure Limit.

Total pressure: The sum of static pressures and velocity pressures within a duct system.

Toxin: A poison.

Transmission loss: The ratio of the amount of sound energy incident on a structure to the sound energy that is transmitted through the structure.

Transport velocity: Minimum air velocity required to move suspended particles in an air stream.

Tridymite: Colorless form of free silica.

TTS: See temporary threshold shift.

Tuberculosis: A diseased caused by infection with the bacterium *mycobacterium tuberculosis.*

TWA: See time-weighted average and Threhsold Limit Value-Time-Weighted Average.

U

UCL: See upper confidence limit.

UEL: See upper explosive limit.

Ultraviolet: Wavelengths of the electromagnetic spectrum shorter than visible light and longer than x-rays.

Unit density: The weight of a particular particle. Unit density is used in particulate sampling to determine the aerodynamic diameter of a particulate and its settling rate.

Upper confidence limit (UCL): A statistical principle applied to estimate the likelihood that a particular value is above the obtained value.

Upper explosive limit (UEL): Upper vapor concentration boundary in which the vapor-air mixture will propagate flame if it is ignited.

V

Vapor: The gaseous form of substances that are normally solid or liquid at room temperature.

Vapor pressure: The pressure exerted by a vapor, measured in pounds per square inch.

Vector: An organism, such as an insect, that transmits pathogens.

Velocity: The derivative of position with respect to time. Also known as speed.

Velocity pressure: The kinetic pressure in the direction of flow necessary to cause a fluid at rest to flow at a given velocity.

Velometer: A device used in ventilation to measure air velocity.

Ventilation: Circulation of air within a room. Local exhaust ventilation removes contaminants at the point of generation and general or dilution ventilation dilutes contaminants to acceptable levels by introducing clean outside air.

Viable: Alive.

Virulence: The capacity of a microorganism to produce disease.

Virus: The causative agent of an infectious disease. Any of a large group of submicroscopic infective agents that are extremely simple microorganisms or extremely complex molecules.

Visible radiation: The wavelengths of the electromagnetic spectrum between 10^{-4} cm to 10^{-5} cm.

Volume flow rate: The quantity of a fluid flowing per unit of time. Sometimes expressed as volume in cubic feet per minute.

W

Watt (W): A unit of power equal to one joule per second.

Wavelength: The distance in the line of advance of a wave from any one point to the next point of corresponding phase.

Wet bulb globe temperature: Temperature as determined by the wet bulb thermometer or a standard sling psychrometer or its equivalent. Influenced by the evaporation rate of water which, in turn, depends on relative air humidity. A wet bulb thermometer consists of a bulb covered with a cloth saturated with water.

Wheatstone bridge: A device for measuring electrical resistance, consisting of a conductor joining two branches of a circuit.

X

X-ray diffraction: An analytical method that measures the pattern of x-rays diffracted by crystalline materials, such as silica.

X-rays: Highly penetrating ionizing radiation formed by electron bombardment.

Z

Z: Symbol for an atomic number. An element's atomic number is the same as the number of protons found in its nucleus.

Zoonoses: Diseases communicable from animals to humans under natural conditions.

References

Lipfert, Frederick, *Air Pollution and Community Health* (New York: Van Nostrand Reinhold, 1994).

American Conference of Governmental Industrial Hygienists, *Air Sampling Instruments for Evaluation of Atmospheric Contaminants*, 6th edition, (Cincinnati, American Conference of Governmental Industrial Hygienists, 1983).

Cockerham and Shane, Eds., *Basic Environmental Toxicology* (Boca Raton, FL: Lewis Publishers, 1993).

Doull, Klaassen, Amdur, Eds., *Casarett and Doull's Toxicology, The Science of Poisons*, 2nd edition (New York: Macmillan Publishers, 1980).

Lippman and Schlesinger, *Chemical Contamination in the Human Environment* (New York: Oxford University Press, 1979).

Putz-Anderson, Vern, Ed., *Cumulative Trauma Disorder: A Manual for Musculoskeletal Diseases of the Upper Limbs* (Cincinnati, OH: National Institute for Occupational Safety and Health, 1988).

Finucane, Edward, *Definitions, Conversions, and Calculations for Occupational Safety and Health Professionals* (Boca Raton, FL: Lewis Publishers, 1993).

Eastman Kodak Company, *Ergonomic Design for People at Work, Volume 1* (New York: Van Nostrand Reinhold, 1989).

Eastman Kodak Company, *Ergonomic Design for People at Work, Volume 2* (New York: Van Nostrand Reinhold, 1989).

Plog, Niland, Quinlin, Eds., *Fundamentals of Industrial Hygiene,* 4th edition (Itasca, IL: National Safety Council, 1996).

Garrett, Cralley, and Cralley, Eds., *Industrial Hygiene Management* (New York: John Wiley and Sons, 1988).

American Conference of Governmental Industrial Hygienists, *Industrial Ventilation: A Manual for Recommended Practice,* 19th edition (Cincinnati, OH: American Conference of Governmental Industrial Hygienists, 1986).

Berger, Ward, Morrill, Royster, Eds., *Noise and Hearing Conservation Manual,* 4th edition (Fairfax, VA: American Industrial Hygiene Association, 1991).

US Department of Labor Occupational Safety and Health Administration, *OSHA Field Operations Manual*, 6th edition (Rockville, MD: Government Institutes, 1994).

US Department of Labor Occupational Safety and Health Administration, *OSHA Technical Manual*, 4th edition (Rockville, MD: Government Institutes, 1996).

Burgess, William, *Recognition of Health Hazards in Industry* (New York: John Wiley and Sons, 1995).

Cralley and Cralley, Eds., *Industrial Hygiene Aspects of Plant Operations, Volume 1, Process Flows* (New York: Macmillan Publishing, 1984).

Cralley and Cralley, Eds., *Industrial Hygiene Aspects of Plant Operations, Volume 2, Unit Operations and Product Fabrication* (New York: Macmillan Publishing, 1984).

Cralley, and Cralley, Eds., *Industrial Hygiene Aspects of Plant Operations, Volume 3, Engineering Considerations in Equipment Selection, Layout, and Building Design* (New York: Macmillan Publishing, 1985).

Sax and Lewis, Eds., *Hawley's Condensed Chemical Dictionary,* 11th Edition (New York: Van Nostrand Reinhold Publishers, 1987).

American Conference of Governmental Industrial Hygienists, *1996 TLVs and BEIs Threshold Limit Values for Chemical Substances and Physical Agents Biological Exposure Indices* (Cincinnati, OH: American Conference of Governmental Industrial Hygienists, 1996).

National Institute for Occupational Safety and Health, *The Industrial Environment–It's Evaluation and Control* (Washington, DC: United States Department of Health, Education and Welfare, 1973).

American National Standards Institute, *Practice for Occupational and Educational Eye and Face Protection,* ANSI Z87 (New York: American National Standards Institute, 1989).

Index

A

B

C

D

E

F

G

I

J

K

L

M

N

Q

R

S

T

U

V

W

X

Y

Z